You Can Count on Monsters

The First 100 Numbers and Their Characters

Richard Evan Schwartz

A K Peters, Ltd.
Natick, Massachusetts

Editorial, Sales, and Customer Service Office

A K Peters, Ltd.
5 Commonwealth Road, Suite 2C
Natick, MA 01760
www.akpeters.com

Library of Congress Cataloging-in-Publication Data

Schwartz, Richard Evan.
You can count on monsters : the first 100 numbers and their characters / Richard Evan Schwartz.
p. cm.
ISBN 978-1-56881-578-7 (alk. paper)
1. Factor tables—Juvenile literature. 2. Numbers, Prime—Juvenile literature. 3. Counting—Juvenile literature. I. Title.

QA51.S39 2010
512.9'23–dc22

2009038661

Printed in the United States of America
14 13 12 11 10 9 8 7 6 5 4 3

To Brienne

This book is about the numbers 1 through 100. I wrote the book to teach my daughters about prime numbers and factoring.

The only thing you really need to know in order to enjoy this book is how to multiply whole numbers together, like 2 and 3.

$2 \times 3 = 6$

This means that 2 groups of 3 dots makes 6 dots.

You could also say that 3 groups of 2 dots make 6 dots.

Or you might say that a 2–by–3 grid of dots makes 6 dots.

Another way to write $2 \times 3 = 6$:

This is called a *factor tree.*

$3 \times 5 = 15$

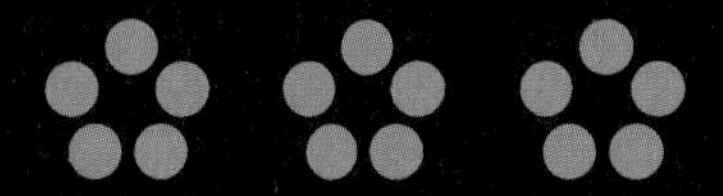

This means that 3 groups
of 5 dots makes 15 dots.

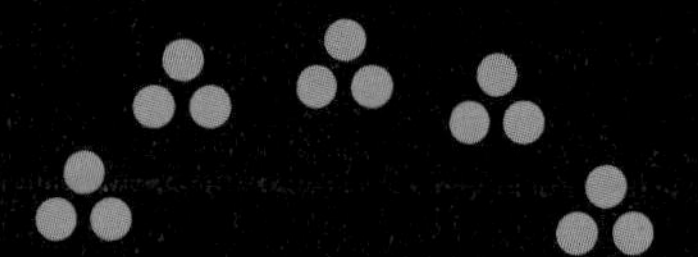

You could also say that 5
groups of 3 dots make 15 dots.

Or you might say that a
3–by–5 grid of dots makes 15 dots.

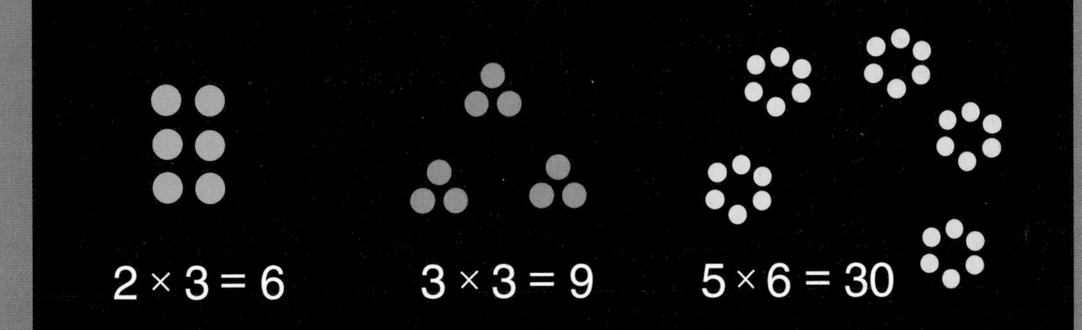

In case you didn't know already, now you know how to multiply whole numbers together. Just arrange some dots into groups of the right size, and then count them.

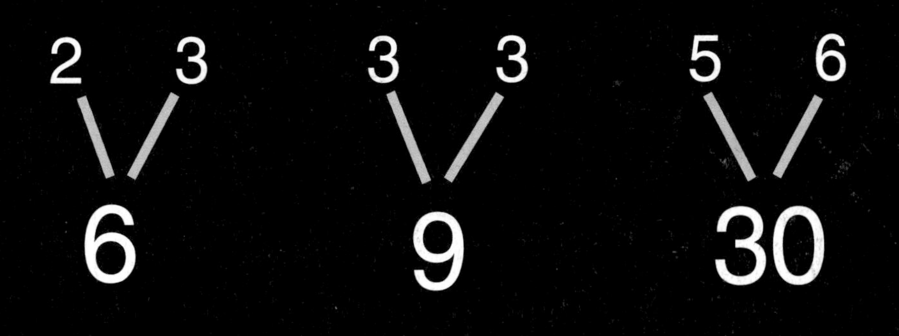

The factor tree we showed for 30 wasn't grown out all the way. Here is a bigger factor tree we get by combining two of the factor trees shown on the previous page.

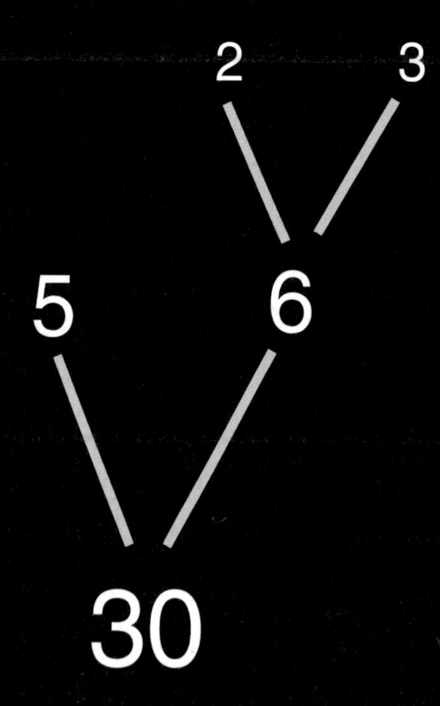

The orange and yellow dots each show how we can arrange 30 dots in a way that reflects the factor tree we have drawn for 30. In the orange case, we've made a 2–by–3 grid of group of 5 dots.

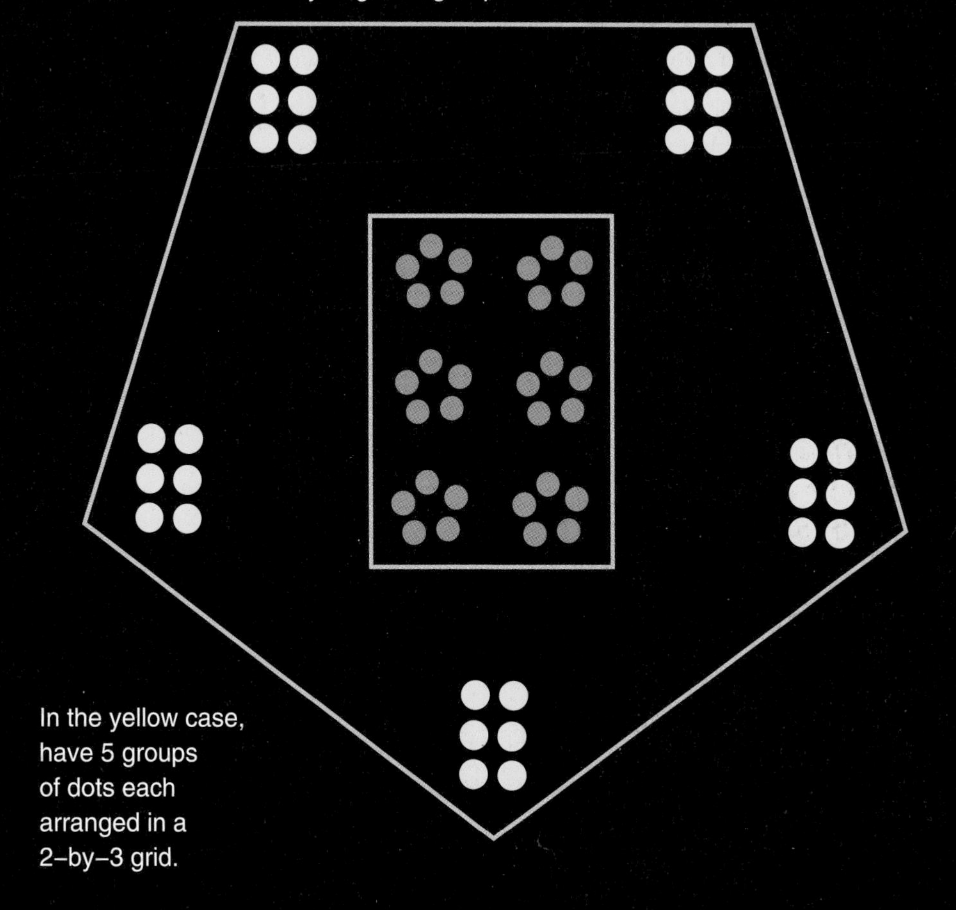

In the yellow case, have 5 groups of dots each arranged in a 2–by–3 grid.

You may wonder if the factor tree we drew for 30 really is all the way grown out. After all, $1 \times 2 = 2$ and $1 \times 3 = 3$ and $1 \times 5 = 5$.

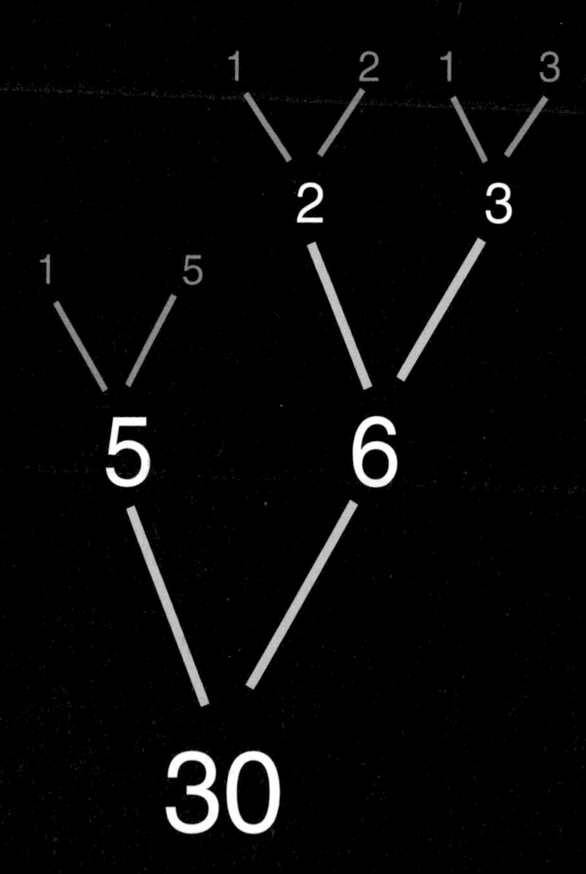

The problem with allowing 1s in the factor tree is that it makes it possible for the tree to grow forever.

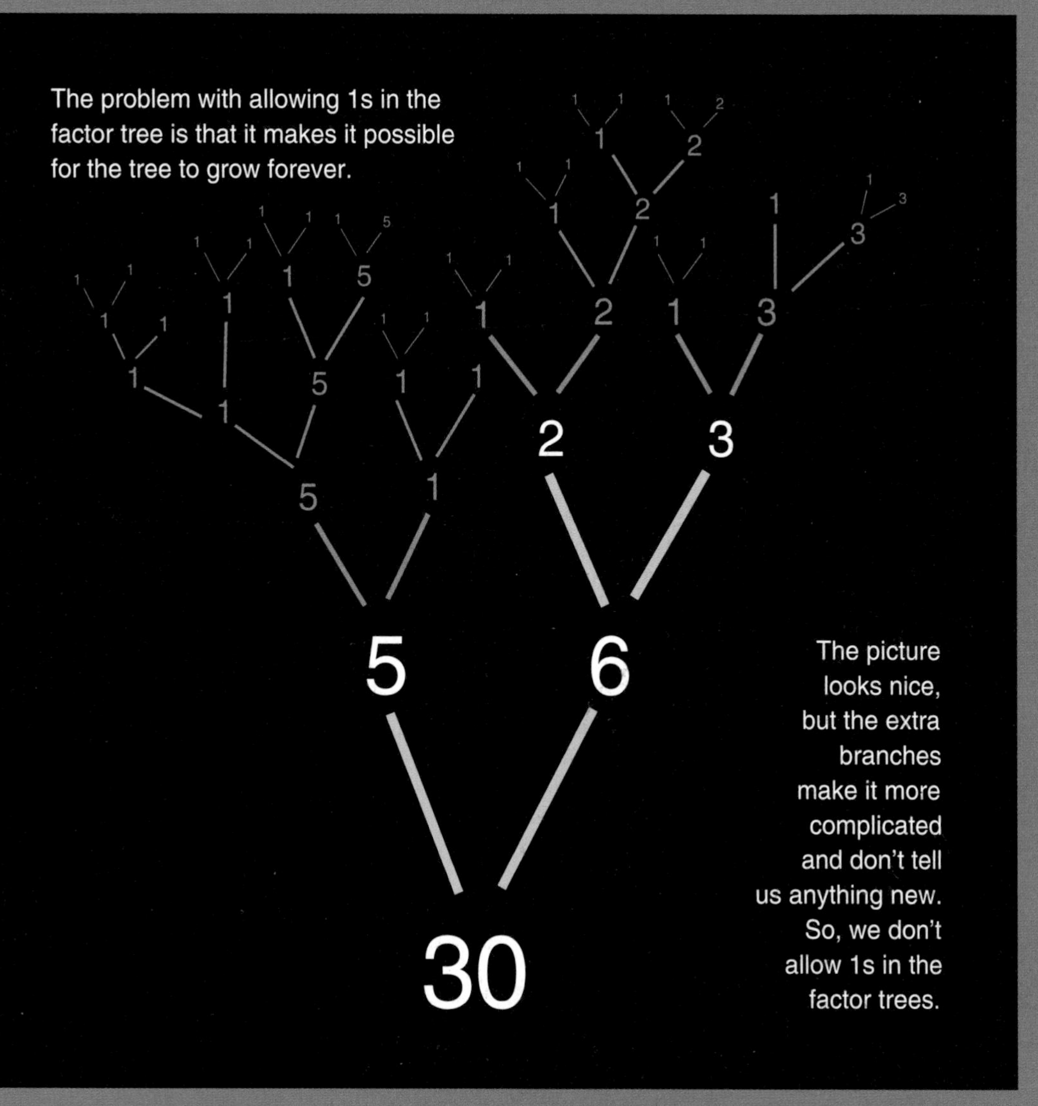

The picture looks nice, but the extra branches make it more complicated and don't tell us anything new. So, we don't allow 1s in the factor trees.

The rules don't mean that you can't draw the crazy kind of tree from the last page. It all depends on the number you choose. With the right number, you can follow the rules *and* draw some crazy trees.

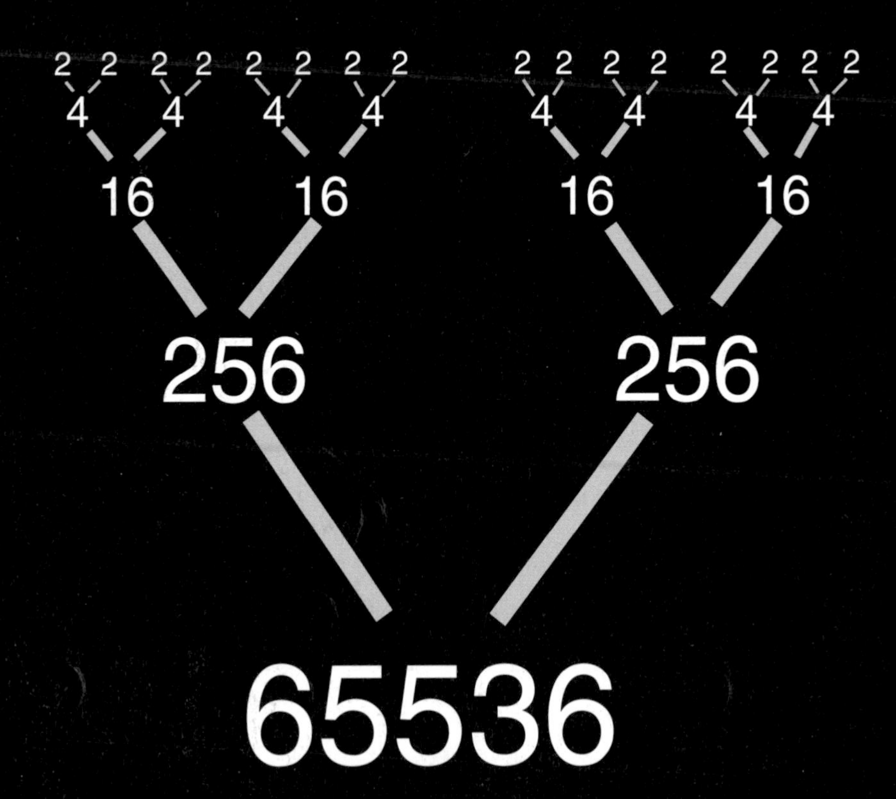

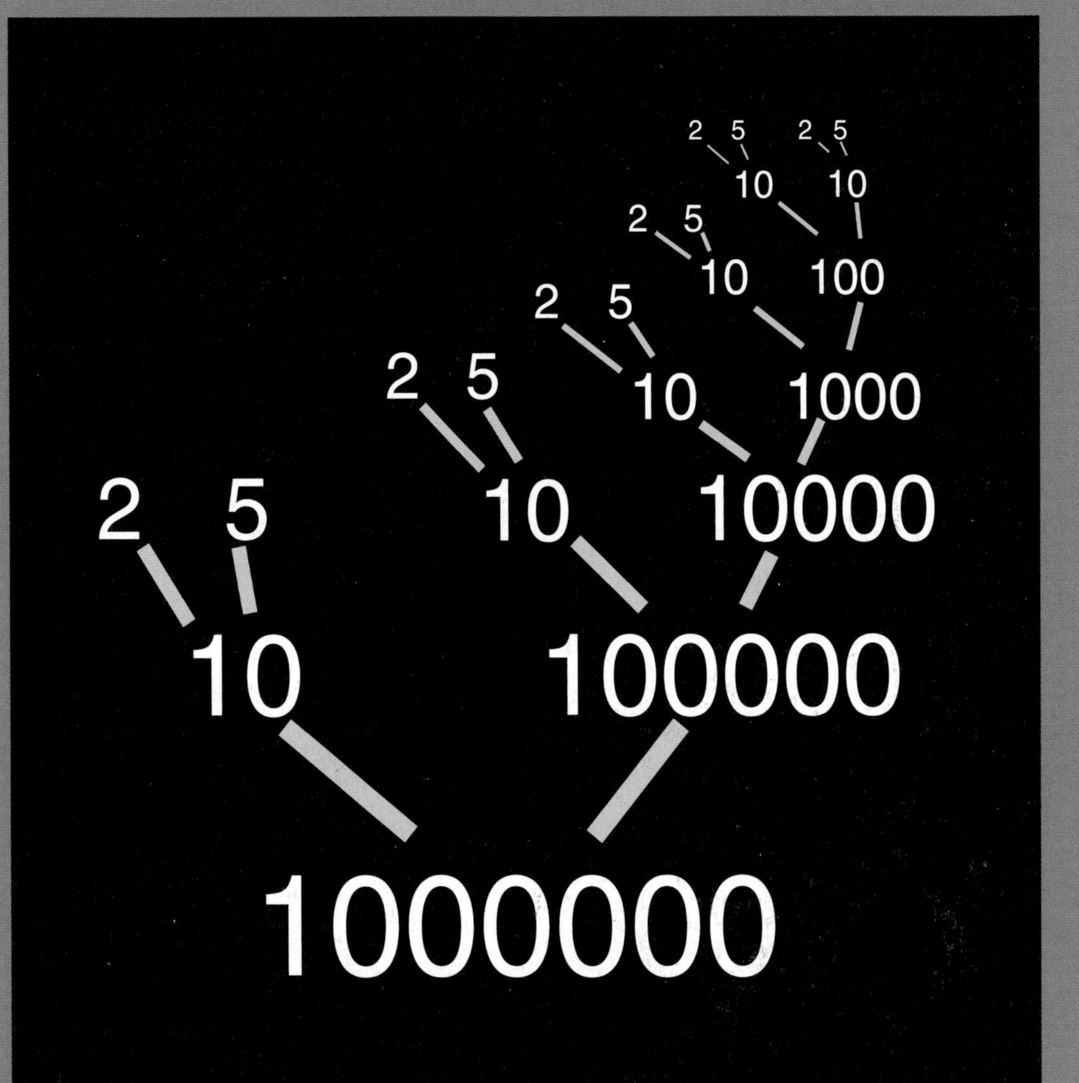

2 5 2 5
10 10
2 5
10 100
2 5
10 1000
2 5
10 10000
2 5
10 100000
1000000

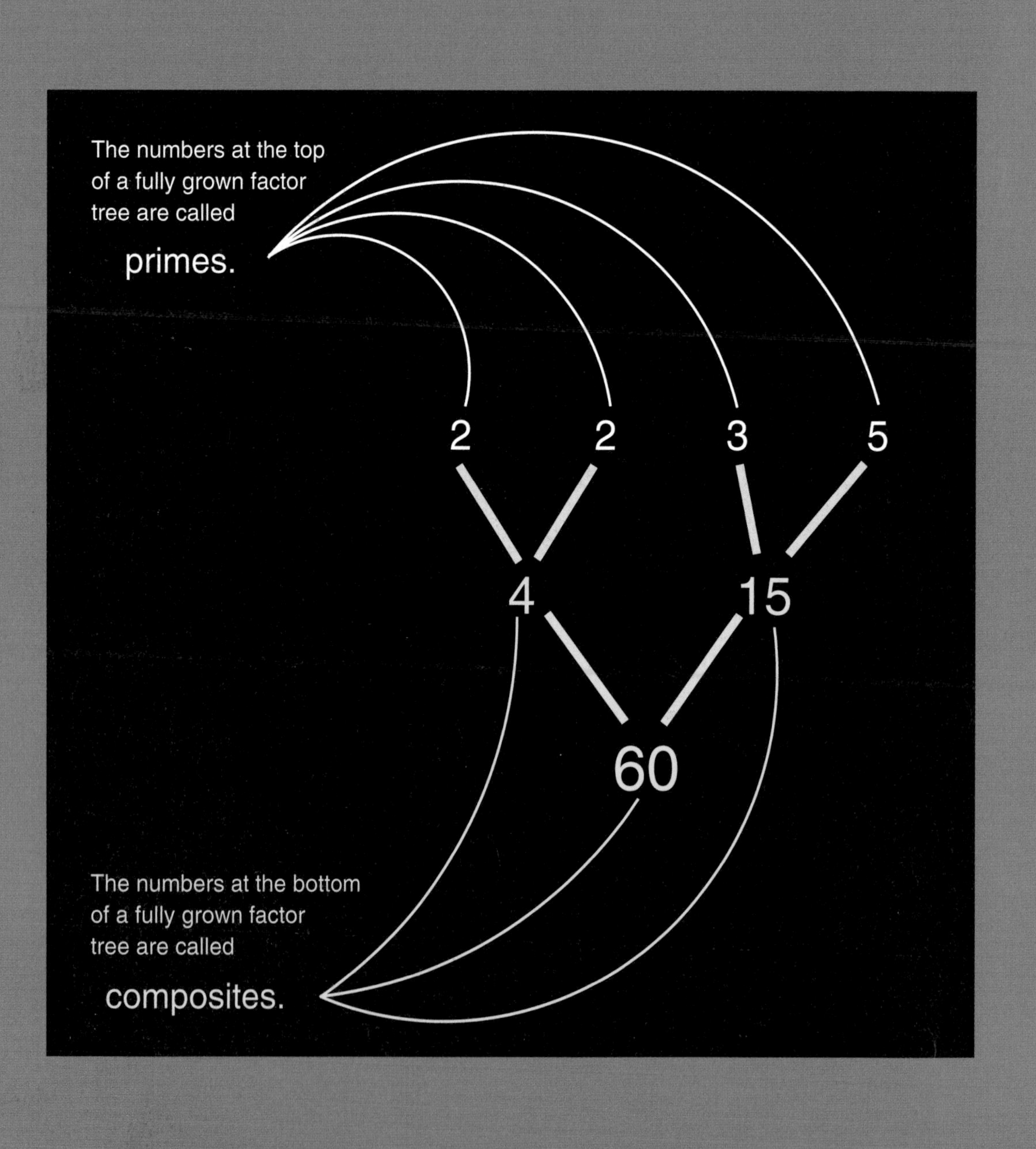
The numbers at the top
of a fully grown factor
tree are called
primes.
2
2
3
5
4
15
60
The numbers at the bottom
of a fully grown factor
tree are called
composites.

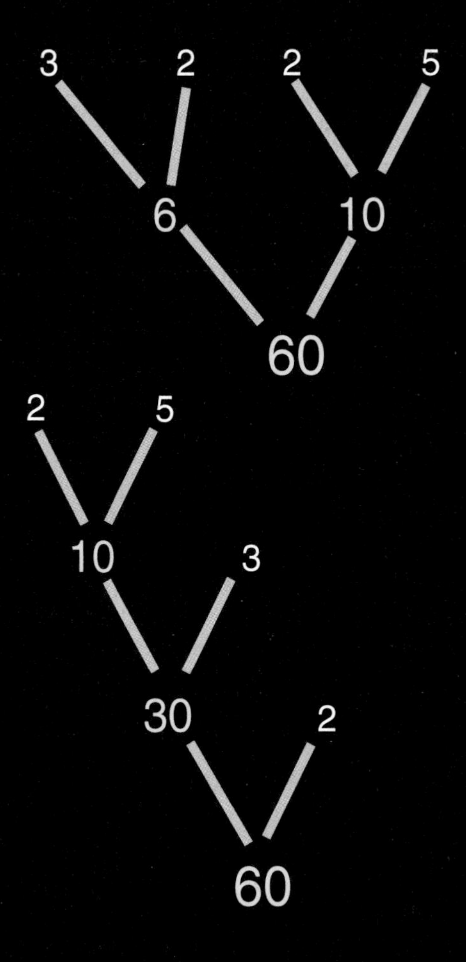

A number usually has many different factor trees, but the primes that appear at the tops of these different trees are all the same, at least when they are listed in order.

These trees all tell us that

$$60 = 2 \times 2 \times 3 \times 5.$$

When you write 60 this way, you are factoring it into primes.

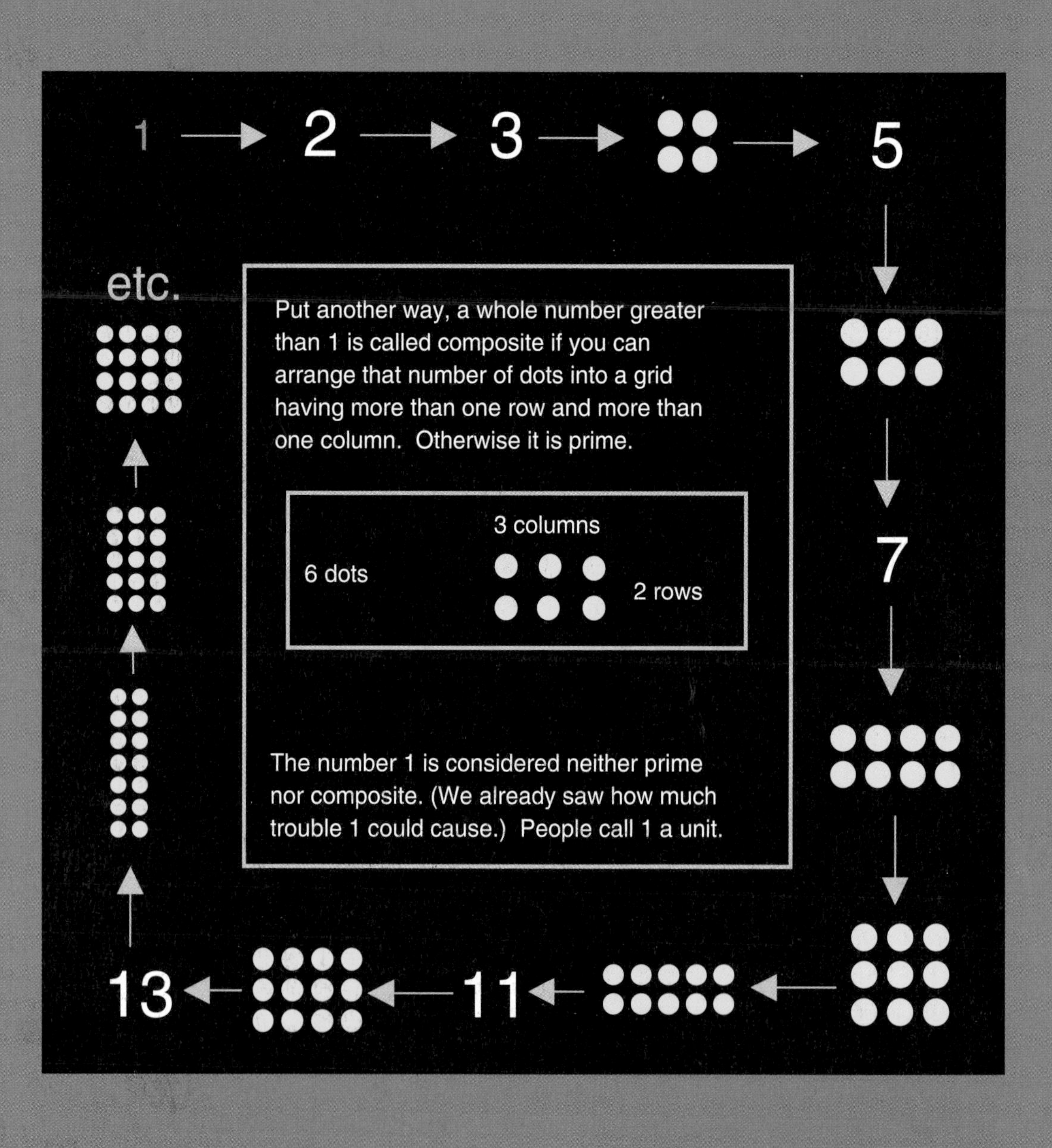
1
2
3
5
7
11
13
etc.
Put another way, a whole number greater than 1 is called composite if you can arrange that number of dots into a grid having more than one row and more than one column. Otherwise it is prime.
3 columns
6 dots
2 rows
The number 1 is considered neither prime nor composite. (We already saw how much trouble 1 could cause.) People call 1 a unit.

The primes less than 100: 2 3 5 7 11 13 17 19 23 29 31 37 41 43 47 53 59 61 67 71 73 79 83 89 97

Every composite can be factored into primes. When it comes to multiplication, the primes are the building blocks of the whole numbers. There are lots of things you can learn about prime numbers. At the end of the book, I'll explain two things you might like to know:

1. How to make a list of all the primes less than your favorite number.

2. Why the primes go on forever.

There are lots of things about primes that have yet to be discovered. For instance, people don't know how to *quickly* factor big composite numbers into primes.

In this book, we will factor the first 100 numbers into primes. For each number, we will show a factor tree and an arrangement of dots that illustrates the nature of the factor tree.

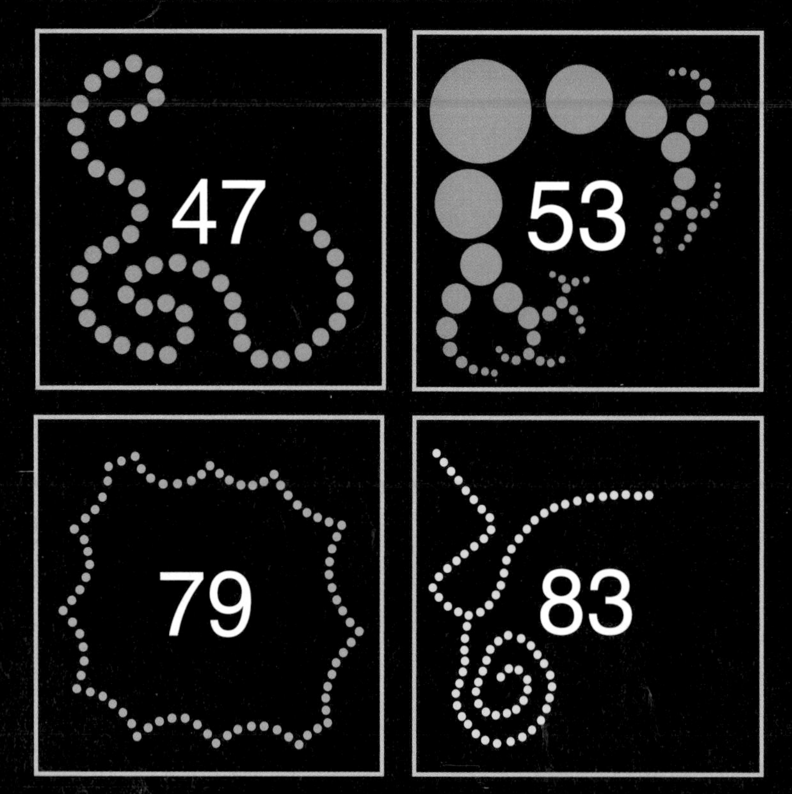

For prime numbers, the factor tree is just a single node, and we will liven up the picture with crazy arrangements of dots. Here are a few of these dot arrangements.

We’re also going to make a monster for each prime number. Here is the 2–monster.

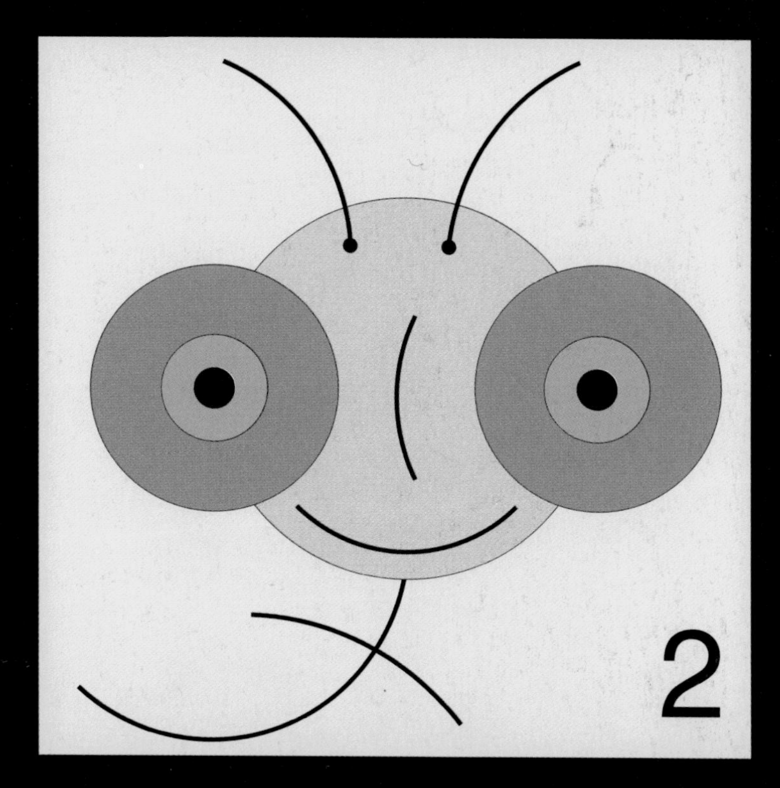

Each monster has something about it that relates to its number, but sometimes you have to look hard (and count) to find it.

Here are the first four prime monsters.

For the composite numbers, we factor the number into primes and then draw a scene that involves the monsters that match those primes.

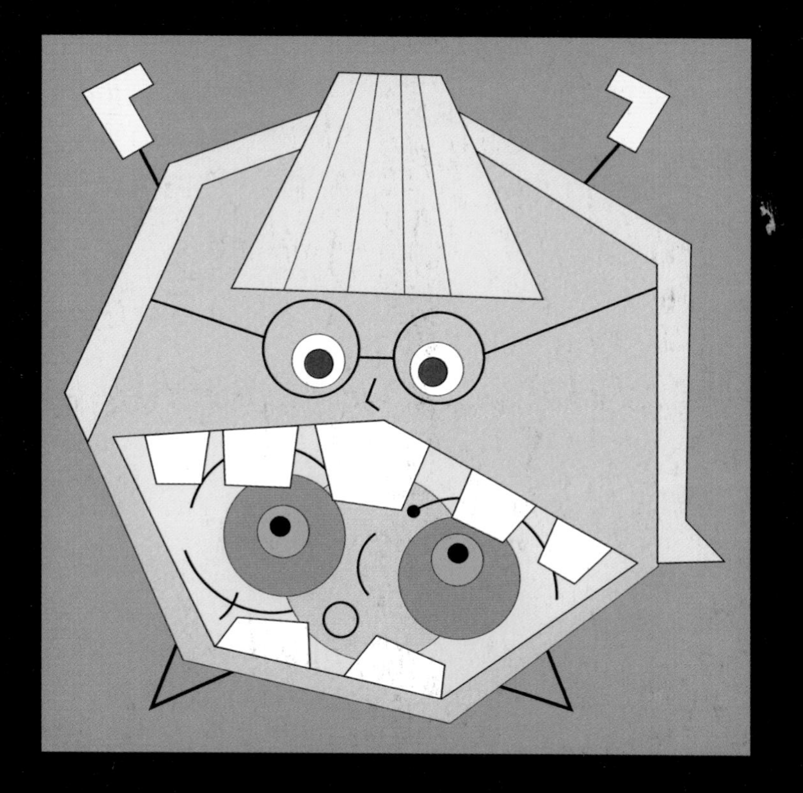

Here is the scene we will draw for the number $14 = 2 \times 7$.

It isn't always easy to recognize the monsters in a scene. For instance, here is the scene for the number 56. You should see three 2–monsters buzzing around one 7–monster.

$$56 = 2 \times 2 \times 2 \times 7$$

Recognizing the monsters in the different scenes is part of the fun of the book!

I imagine that different sides of the monsters are revealed, picture by picture, as they interact with each other.

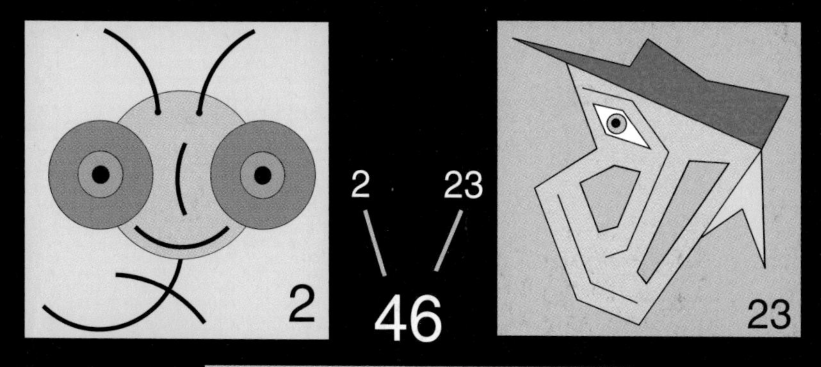

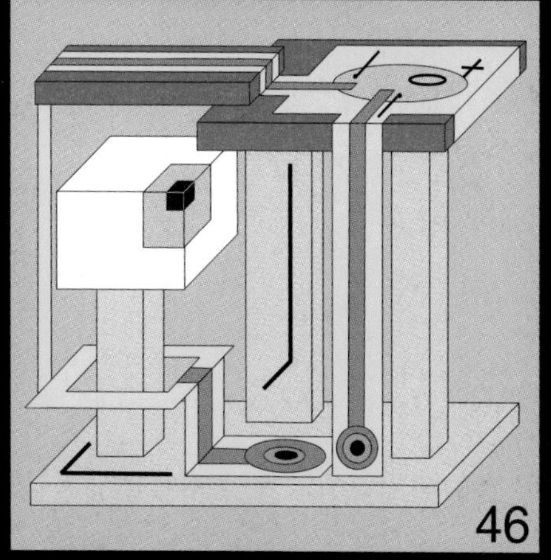

That's how it usually is with people.

Before we get started, there is one more thing that you should know.

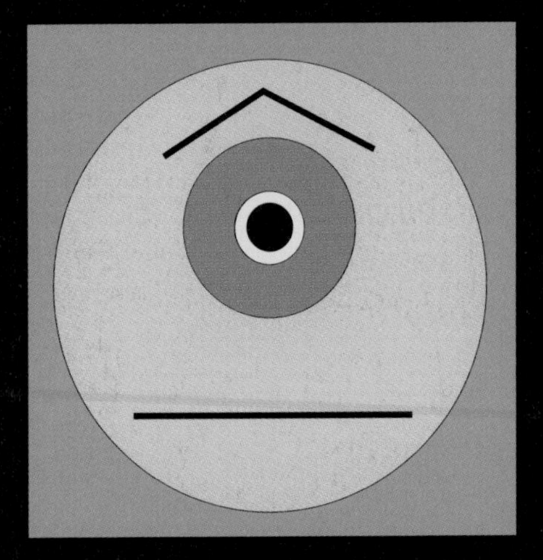

The number 1 gets a monster too, even though 1 is neither prime nor composite. This monster is a bit disappointed because it doesn't get to interact with any of the other monsters.

And here they are!

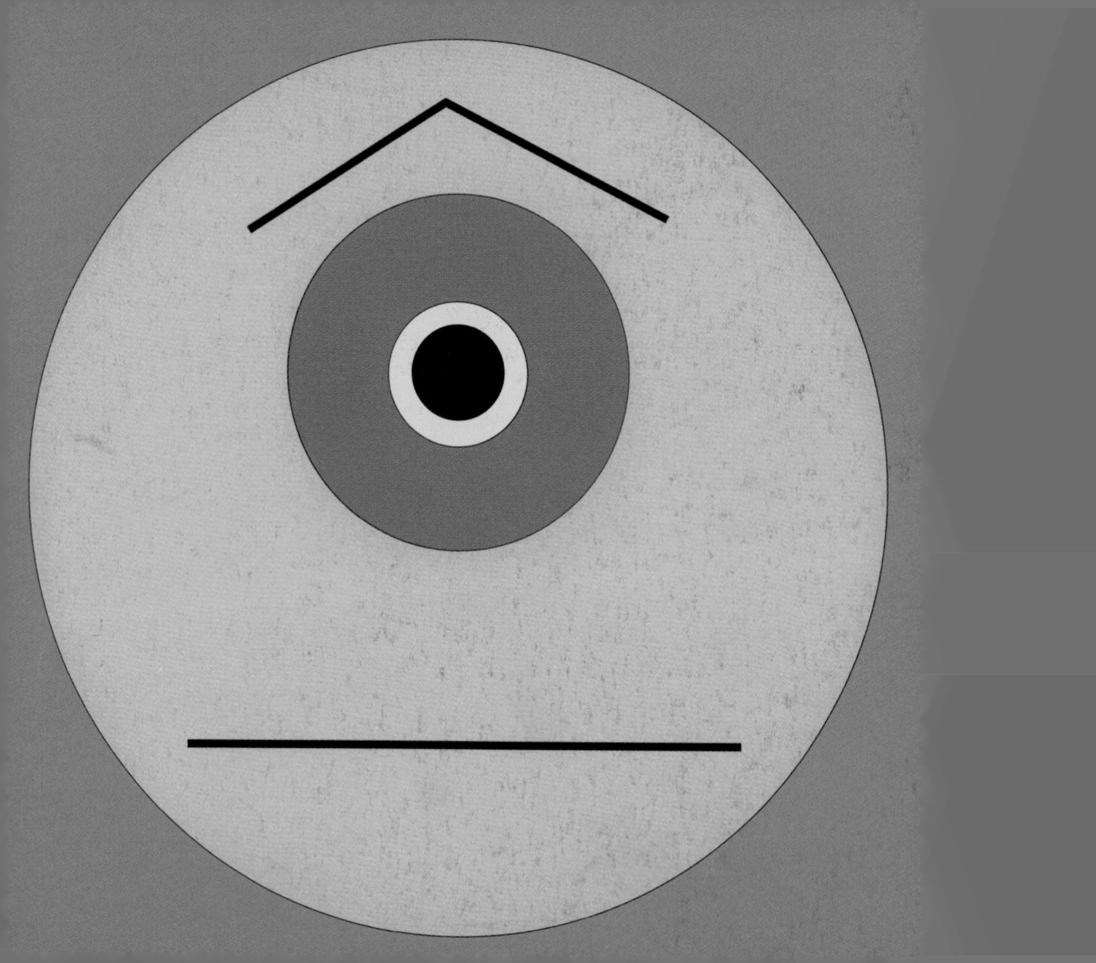

• 2 •

3

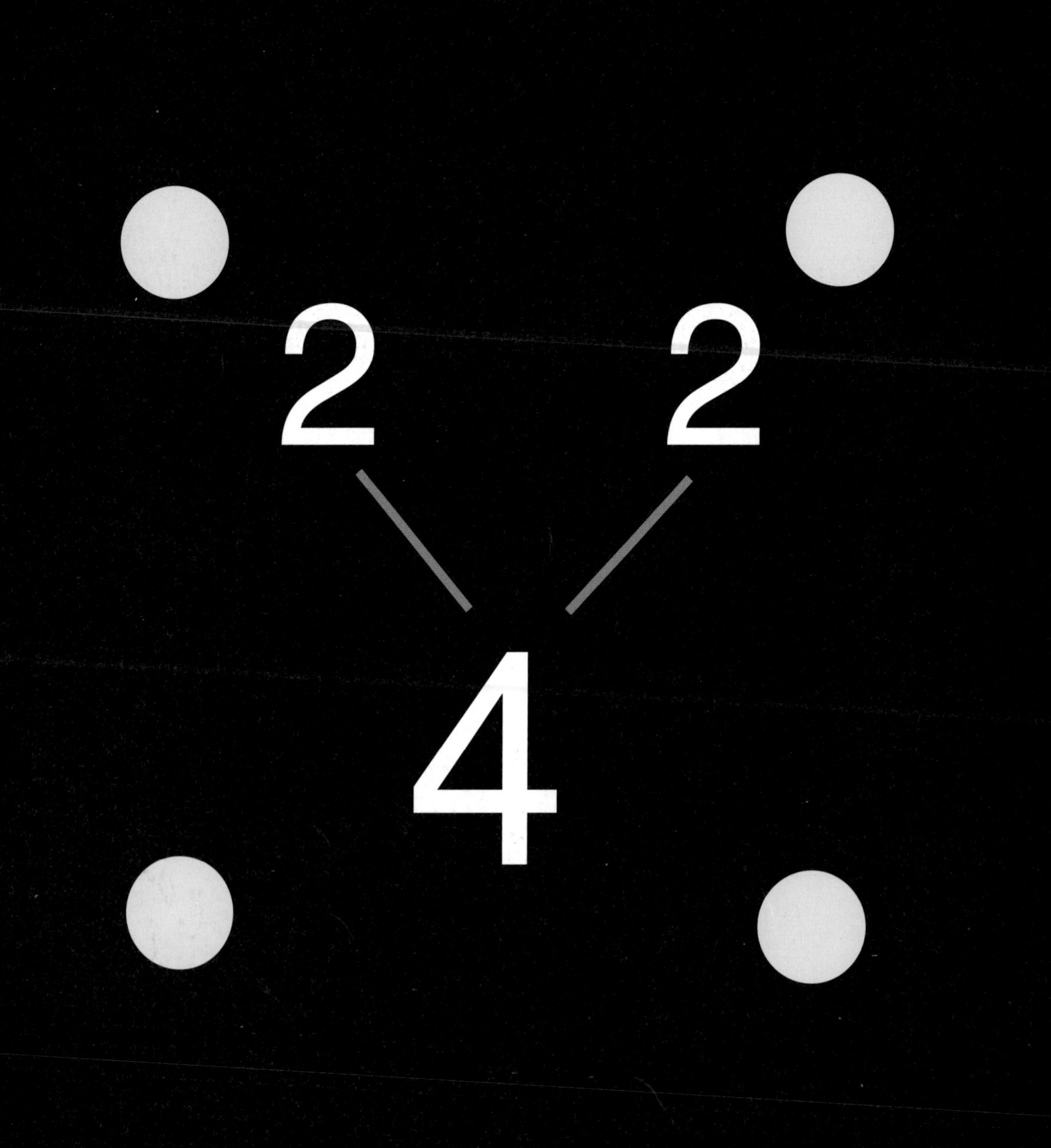
2
2
4

5

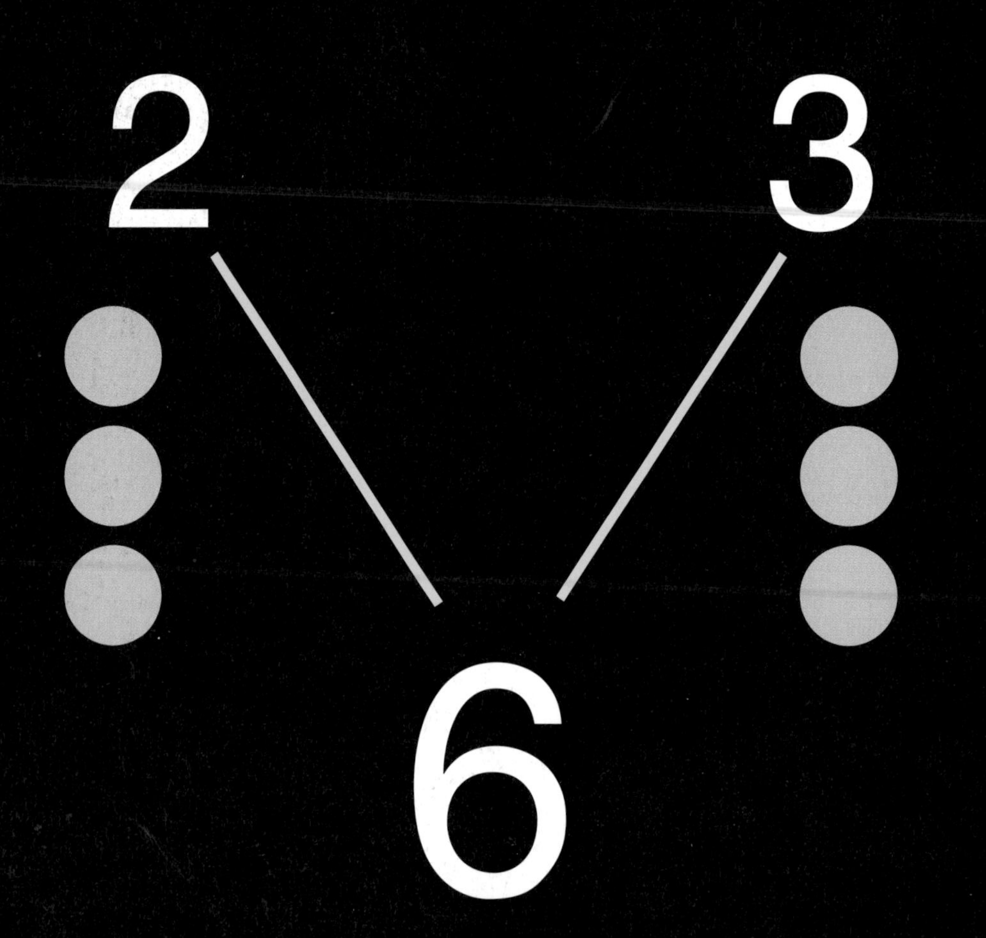
2
3
6

7

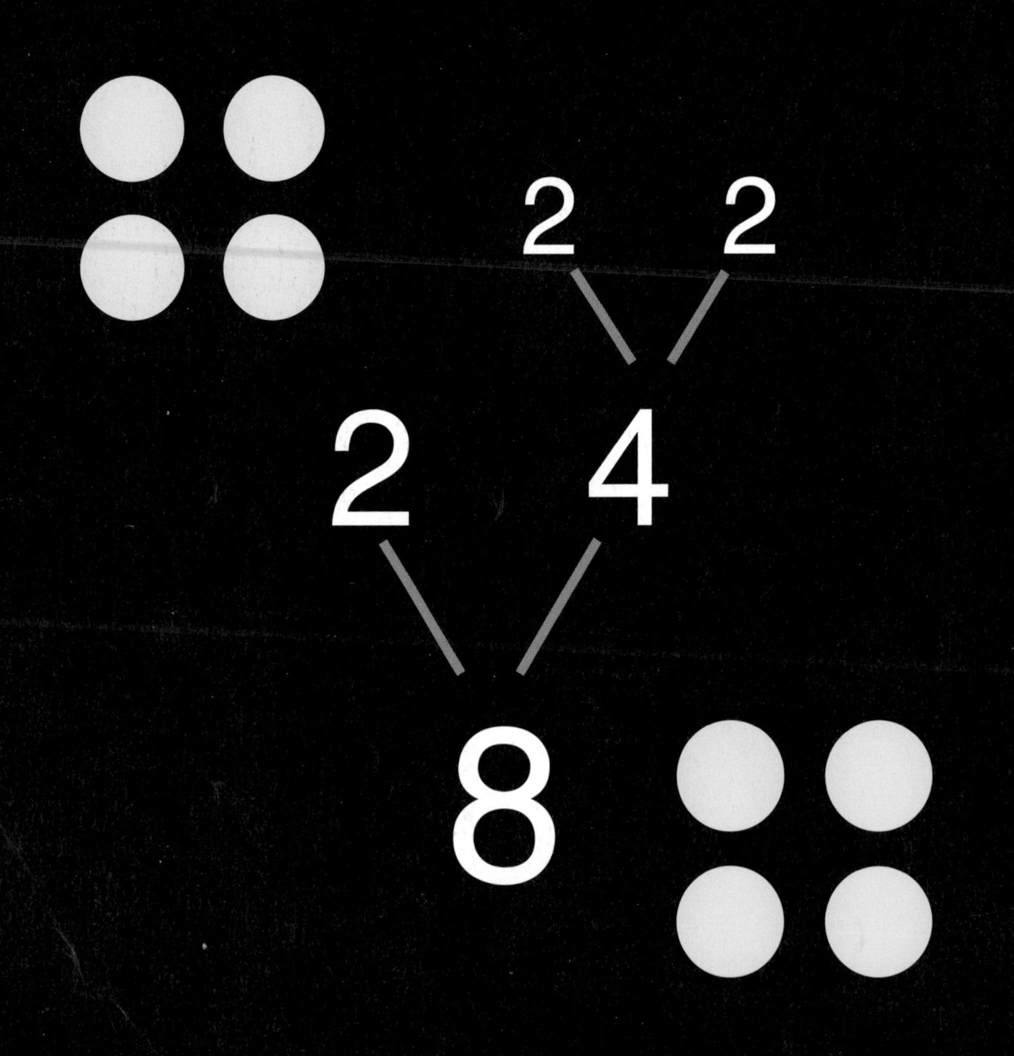
2 2
2 4
8

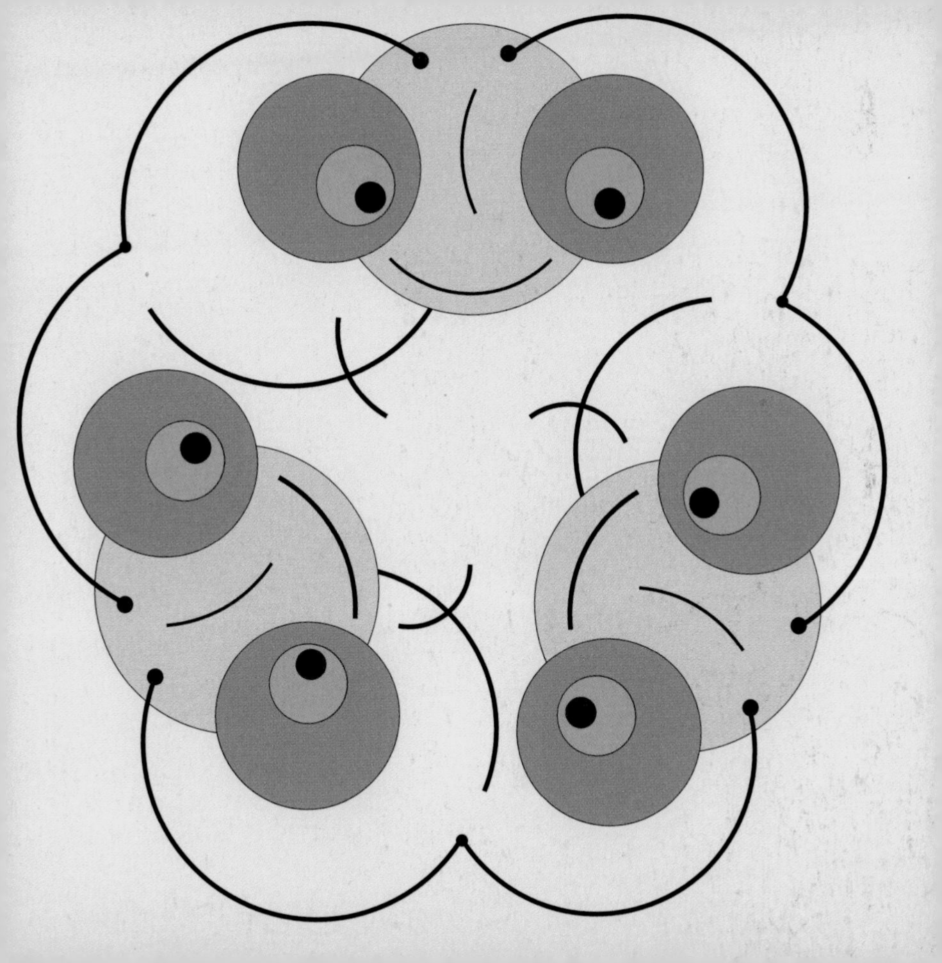

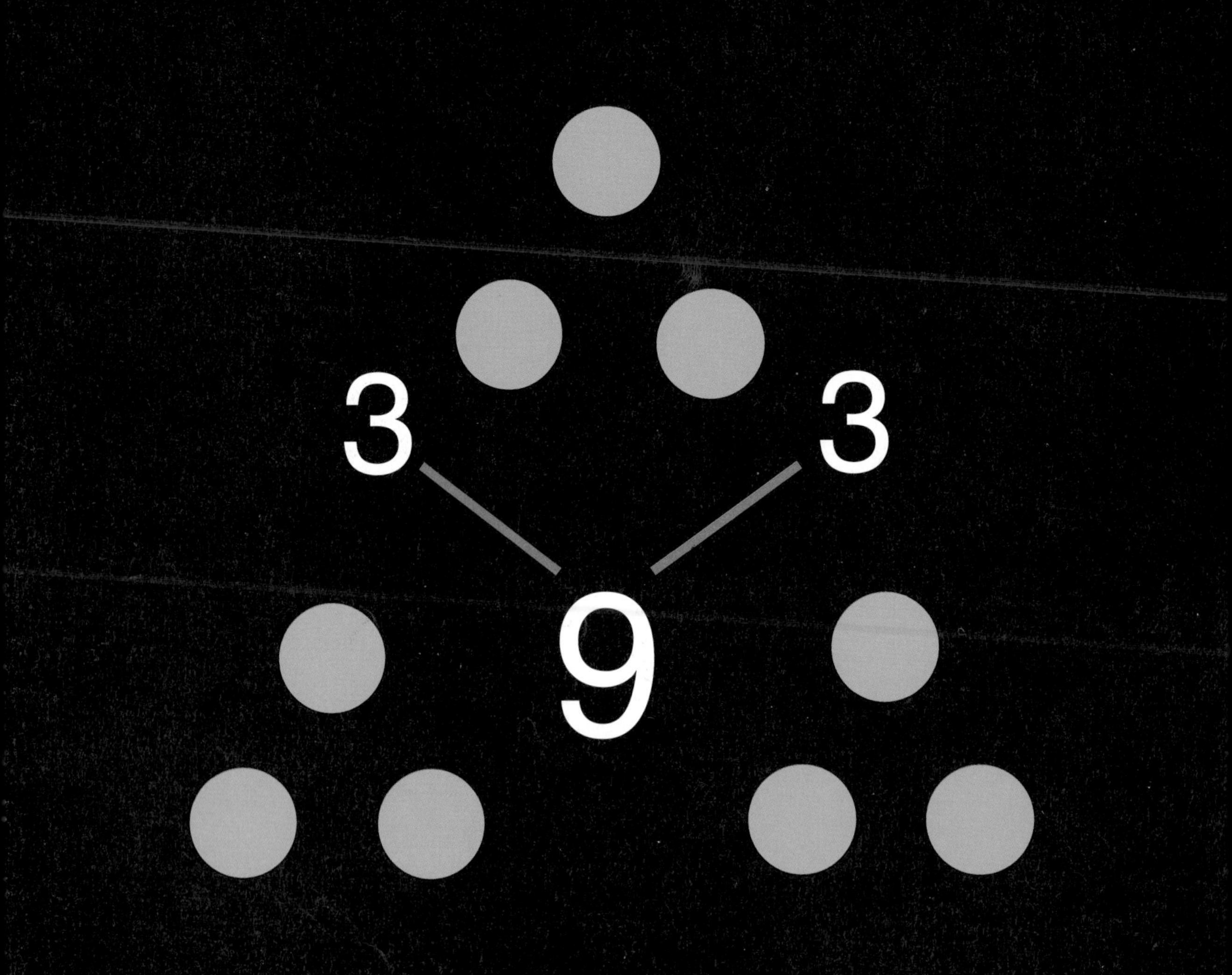
3
3
9

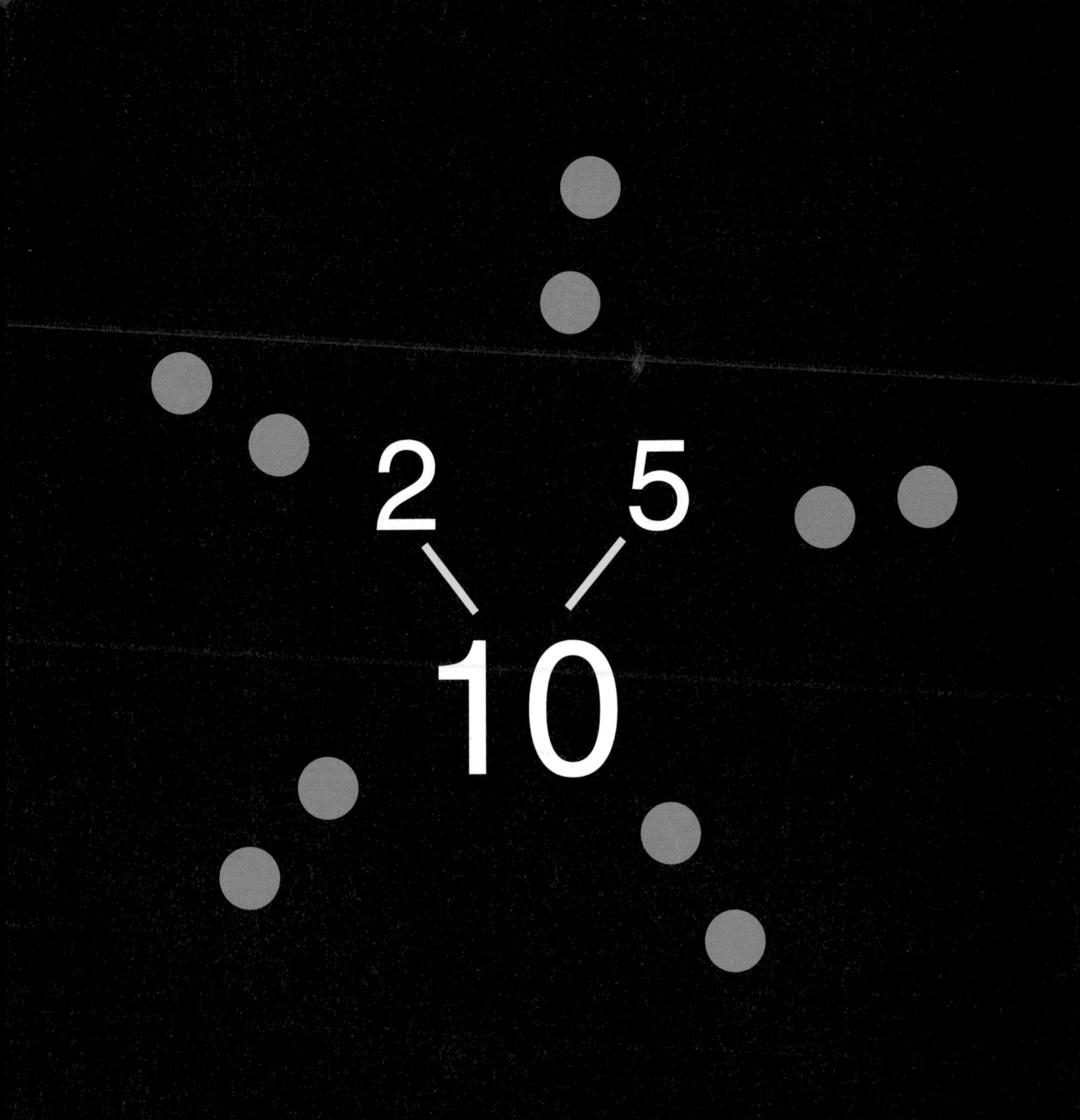
2
5
10

11

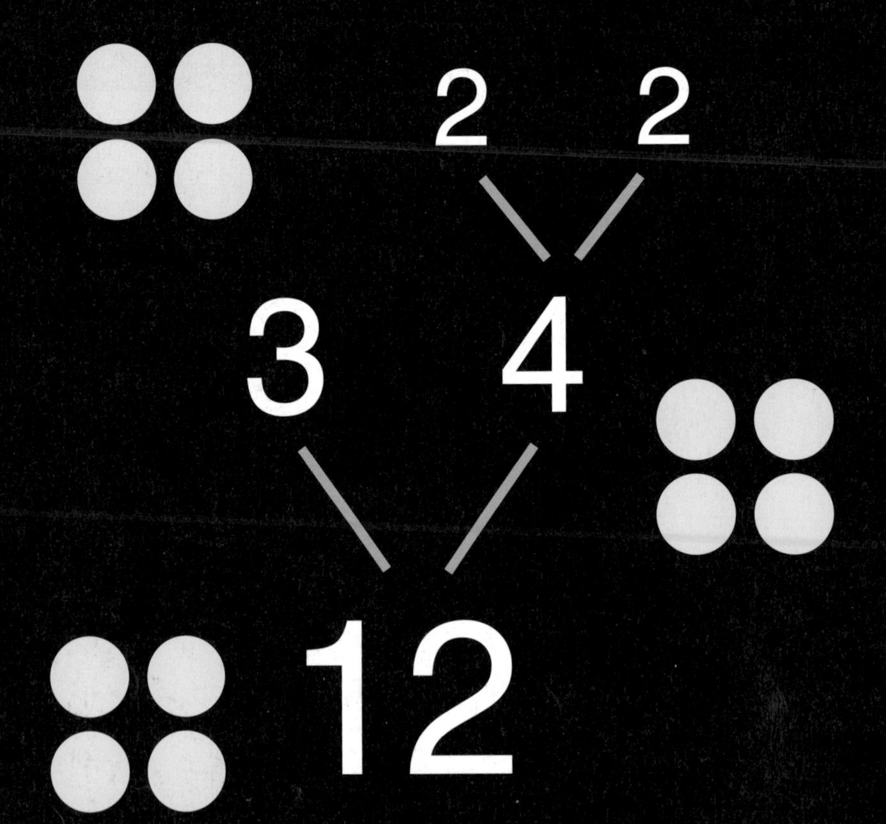
2
2
3
4
12

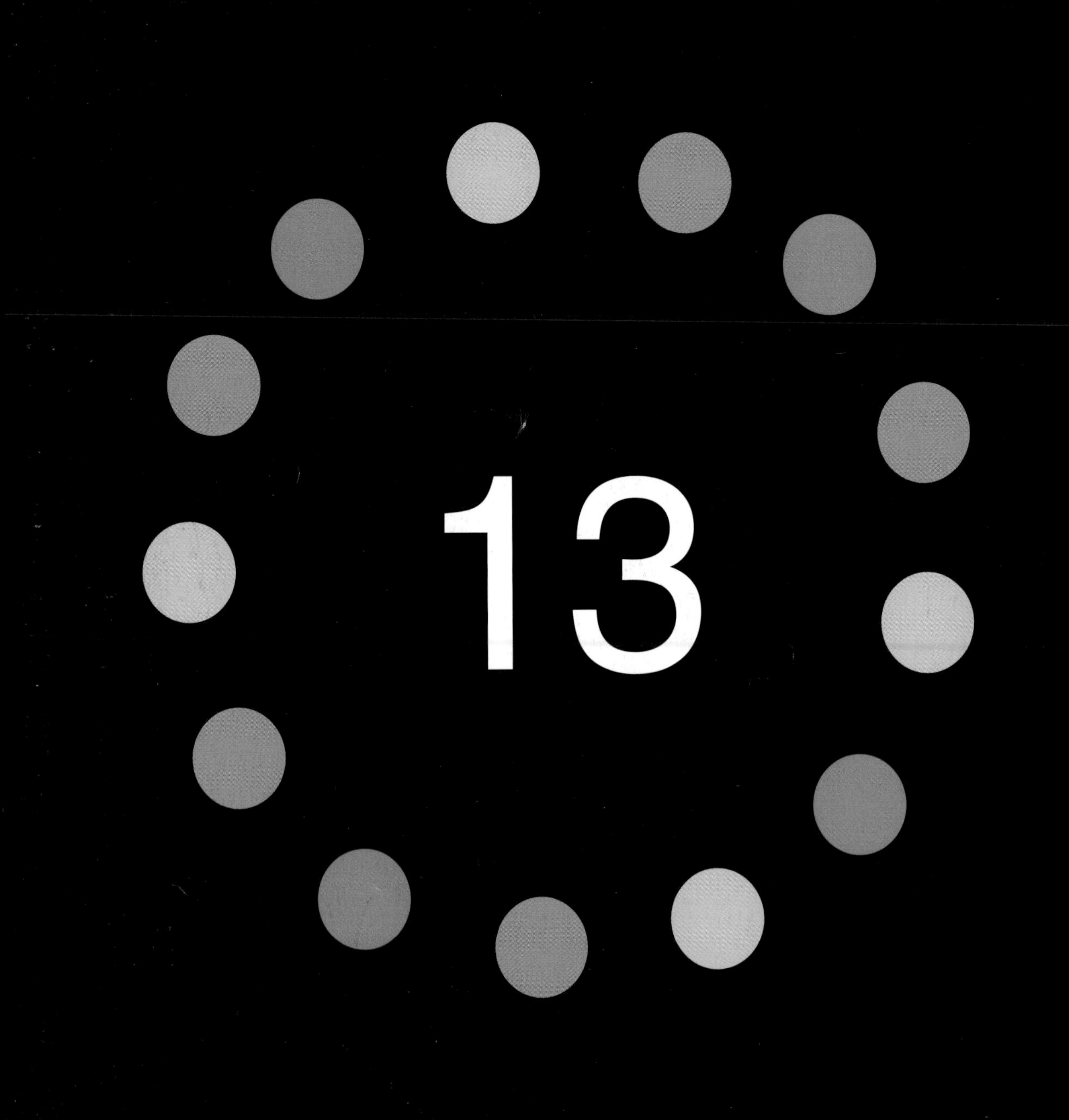
13

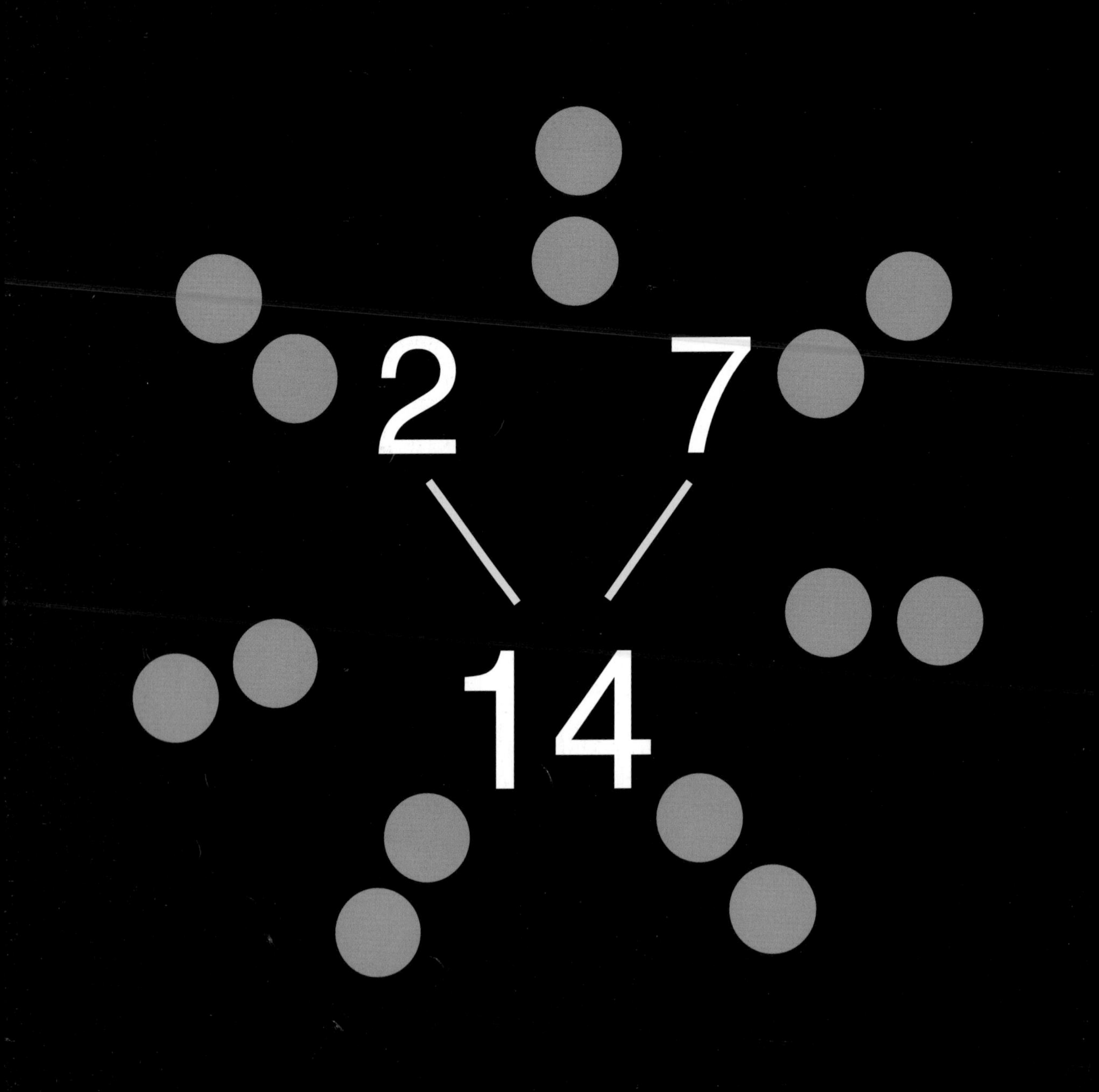

2
7
14

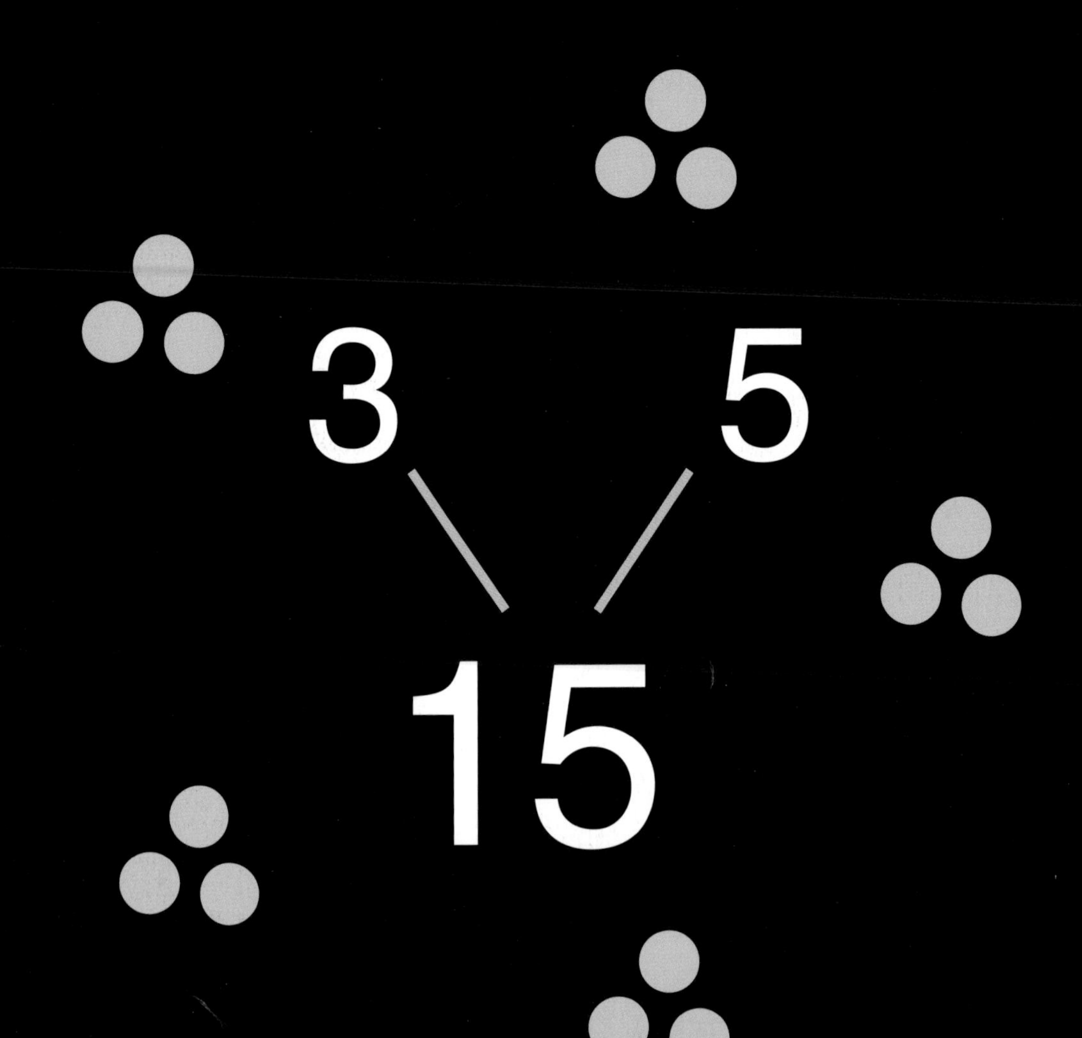
3
5
15

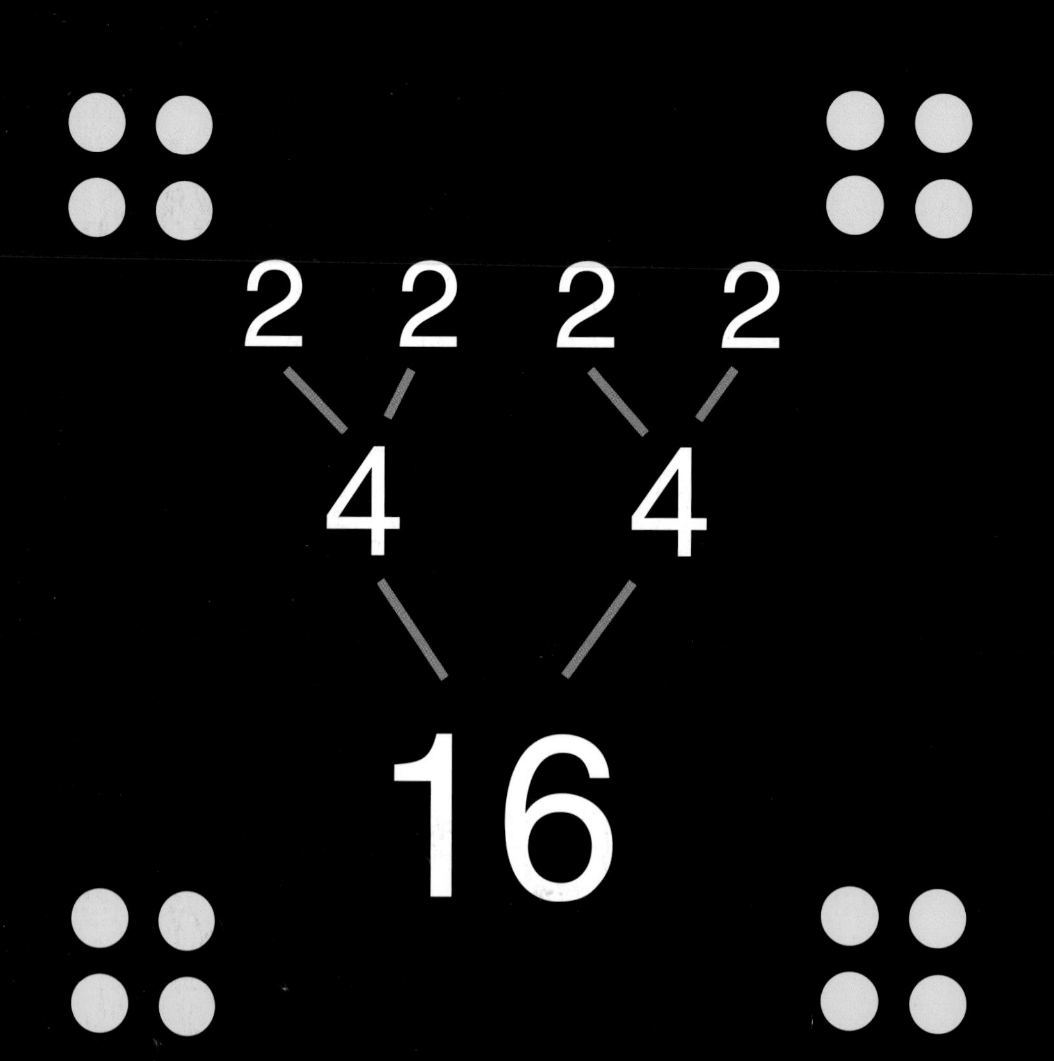
2
2
2
2
4
4
16

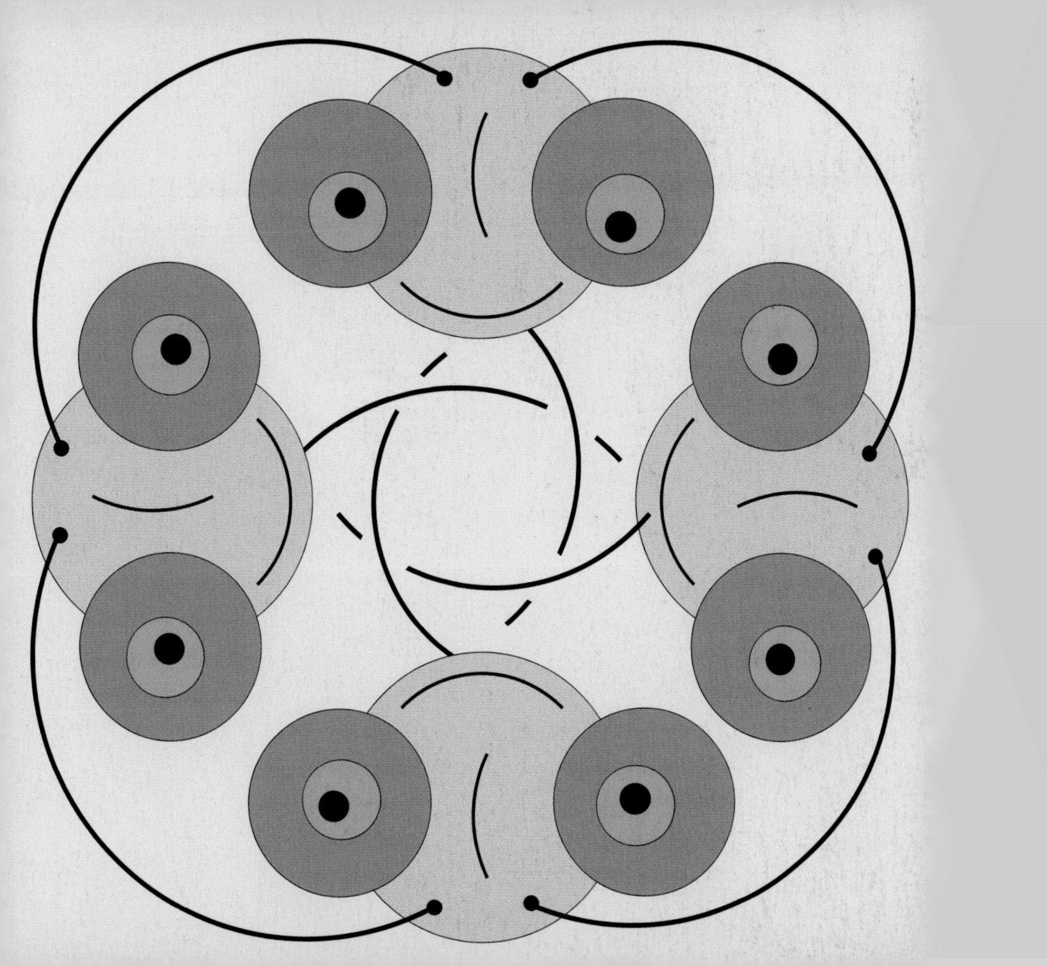

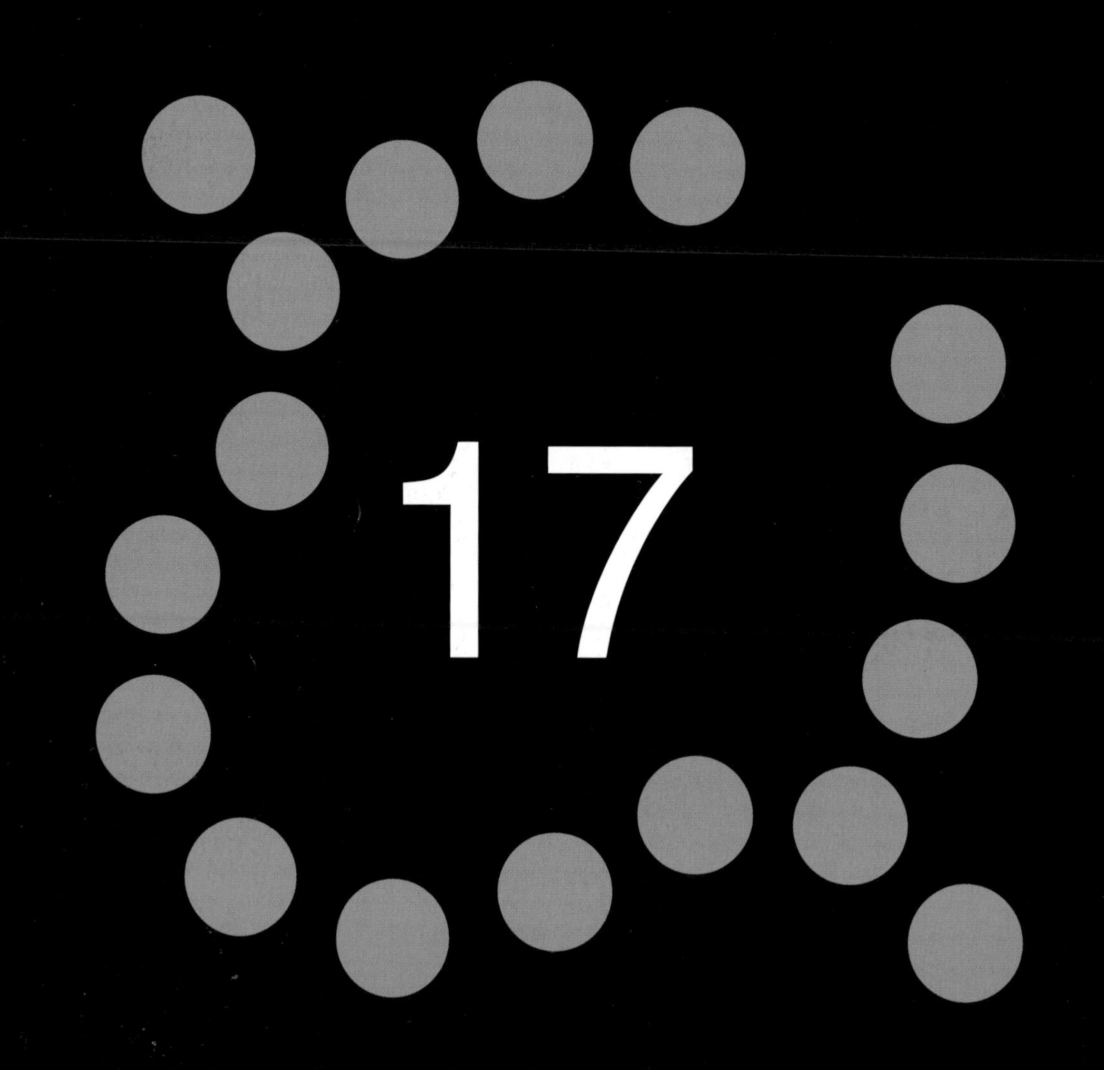
17

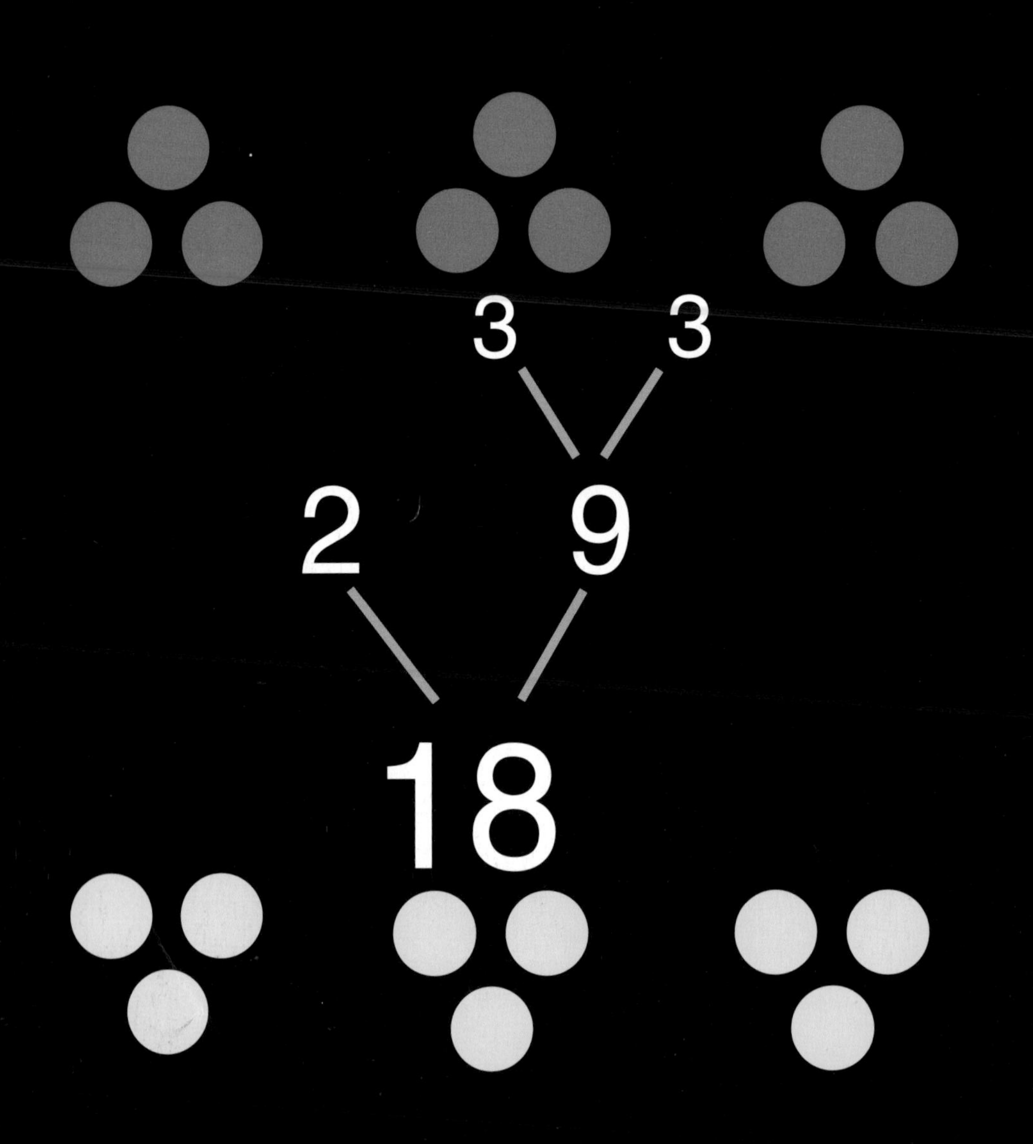
3
3
2
9
18

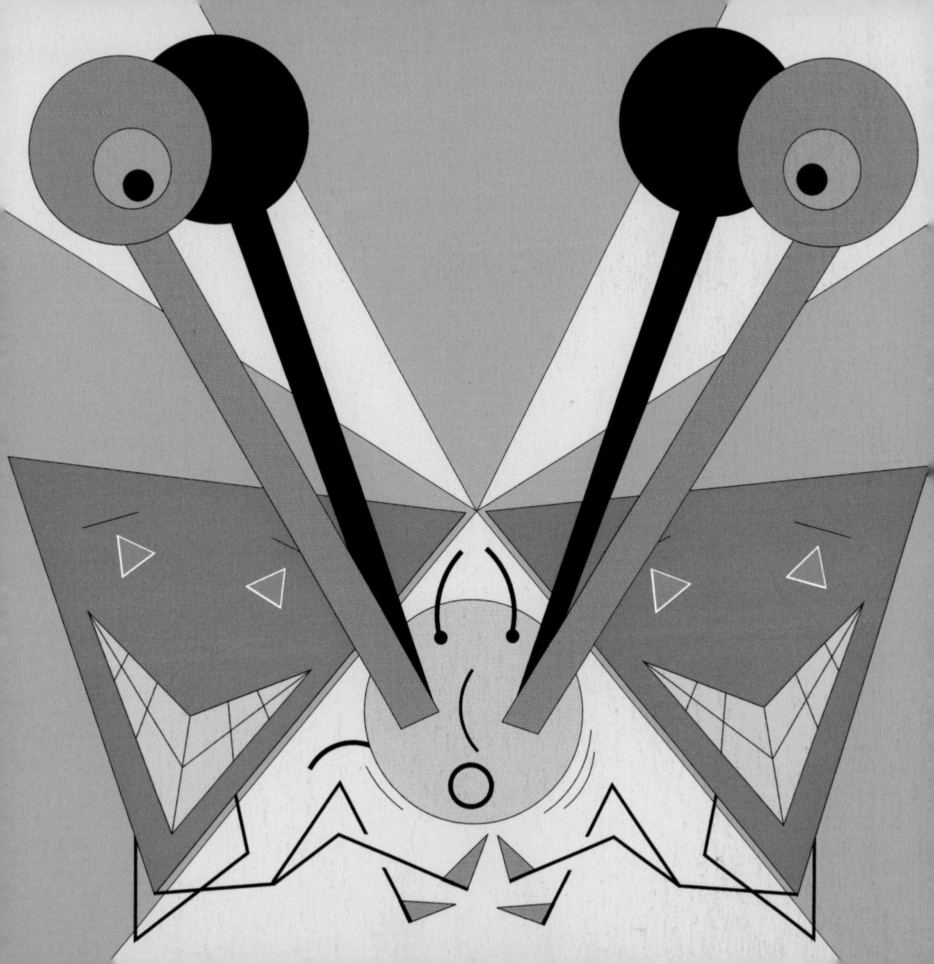

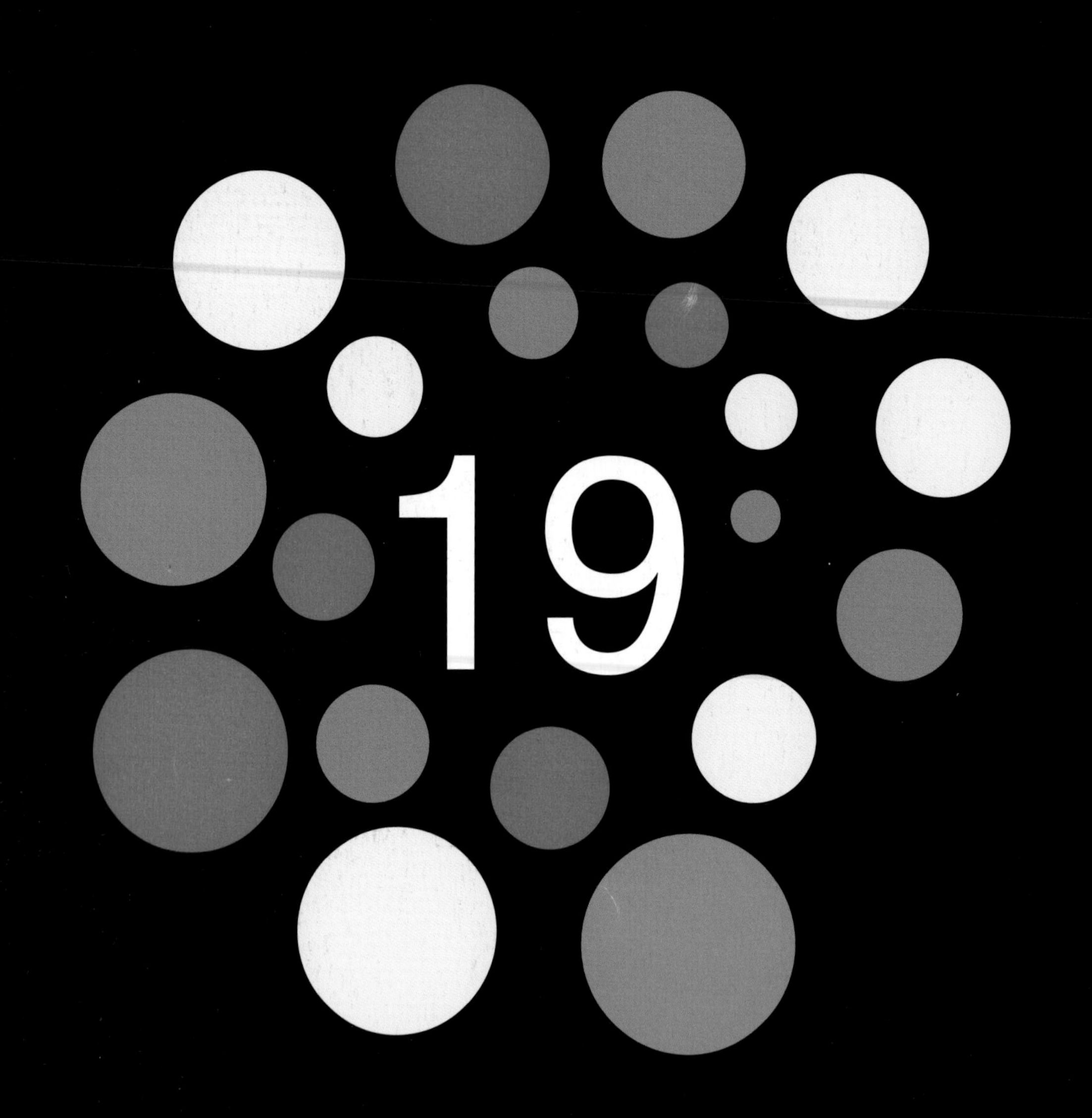
19

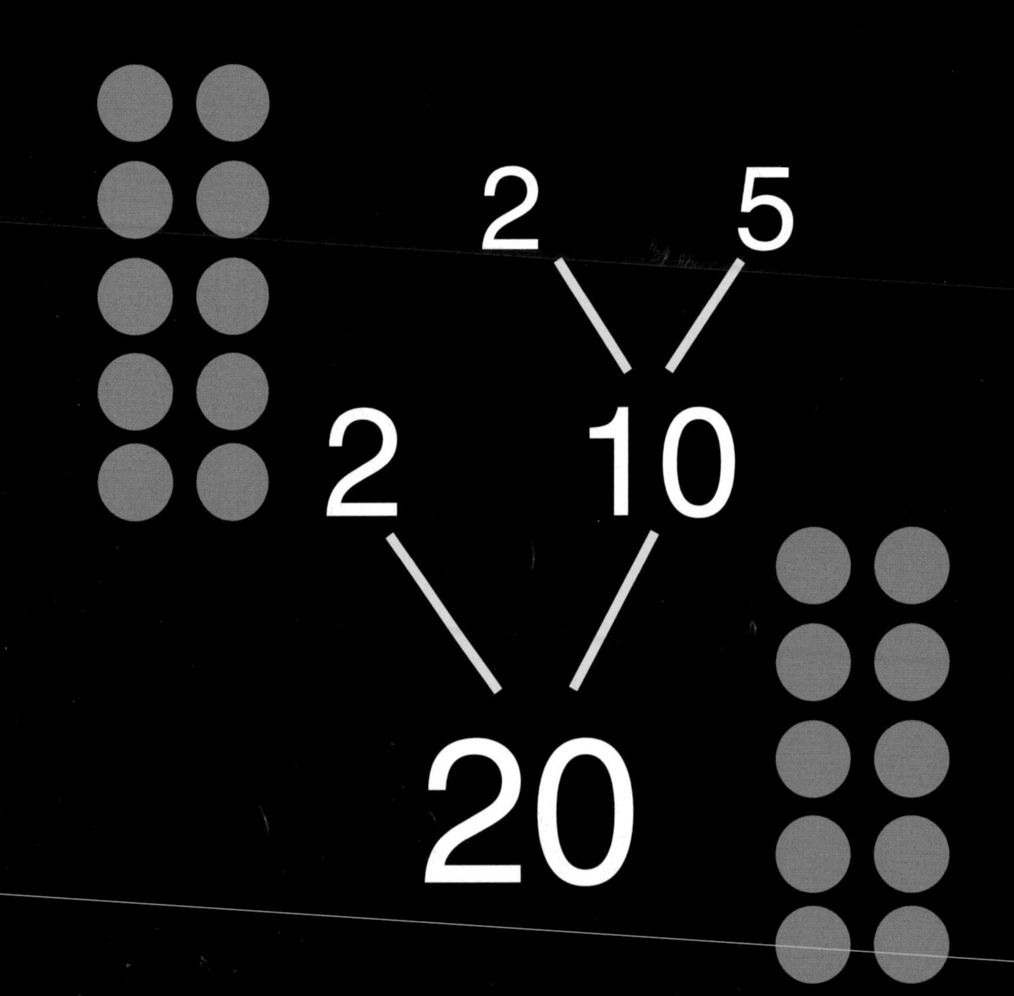
2
5
2
10
20

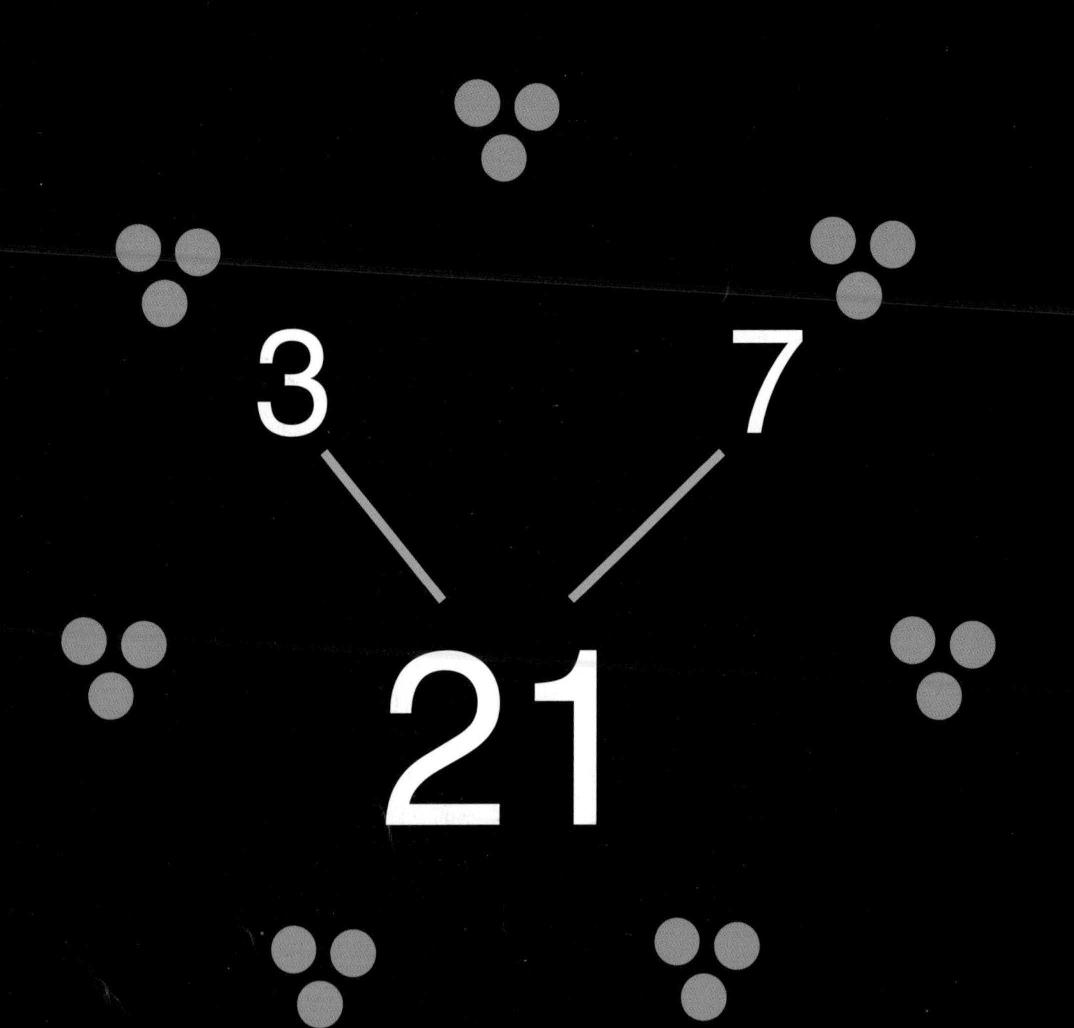
3
7
21

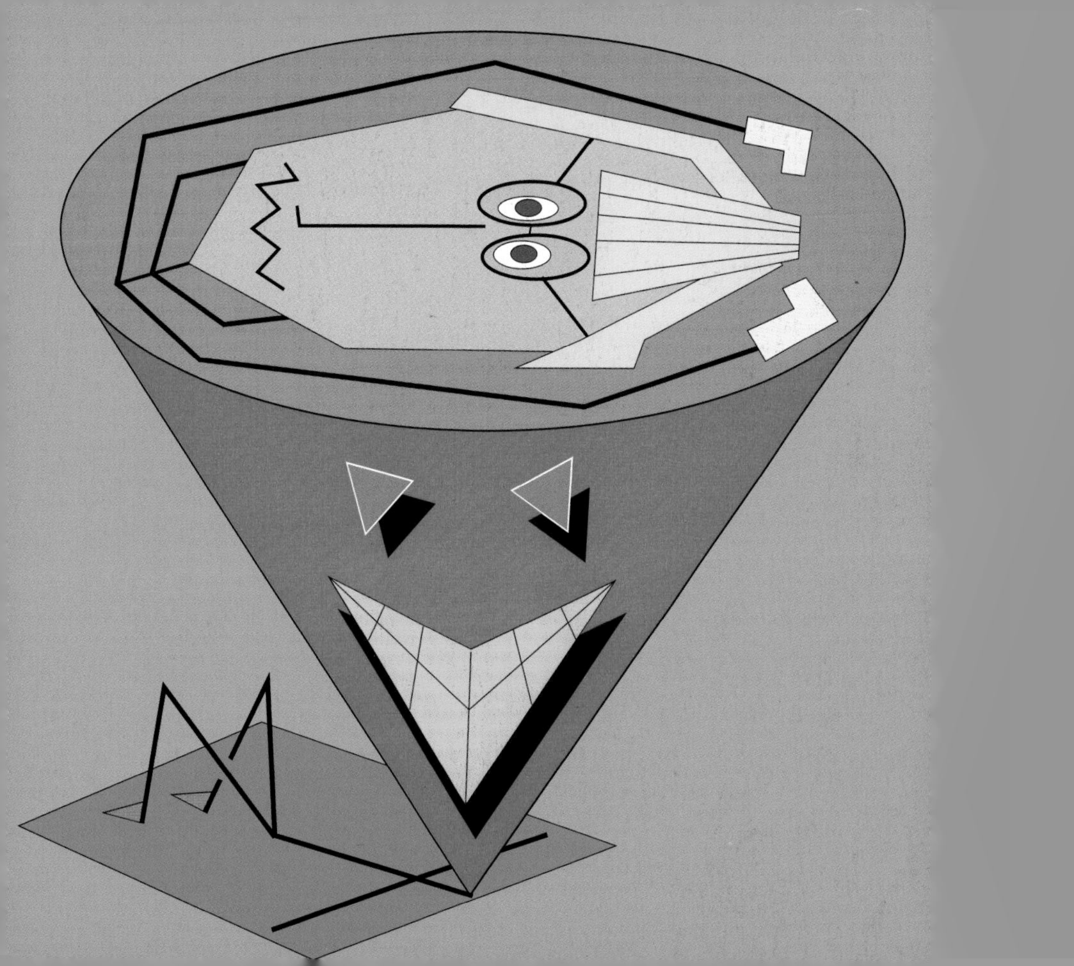

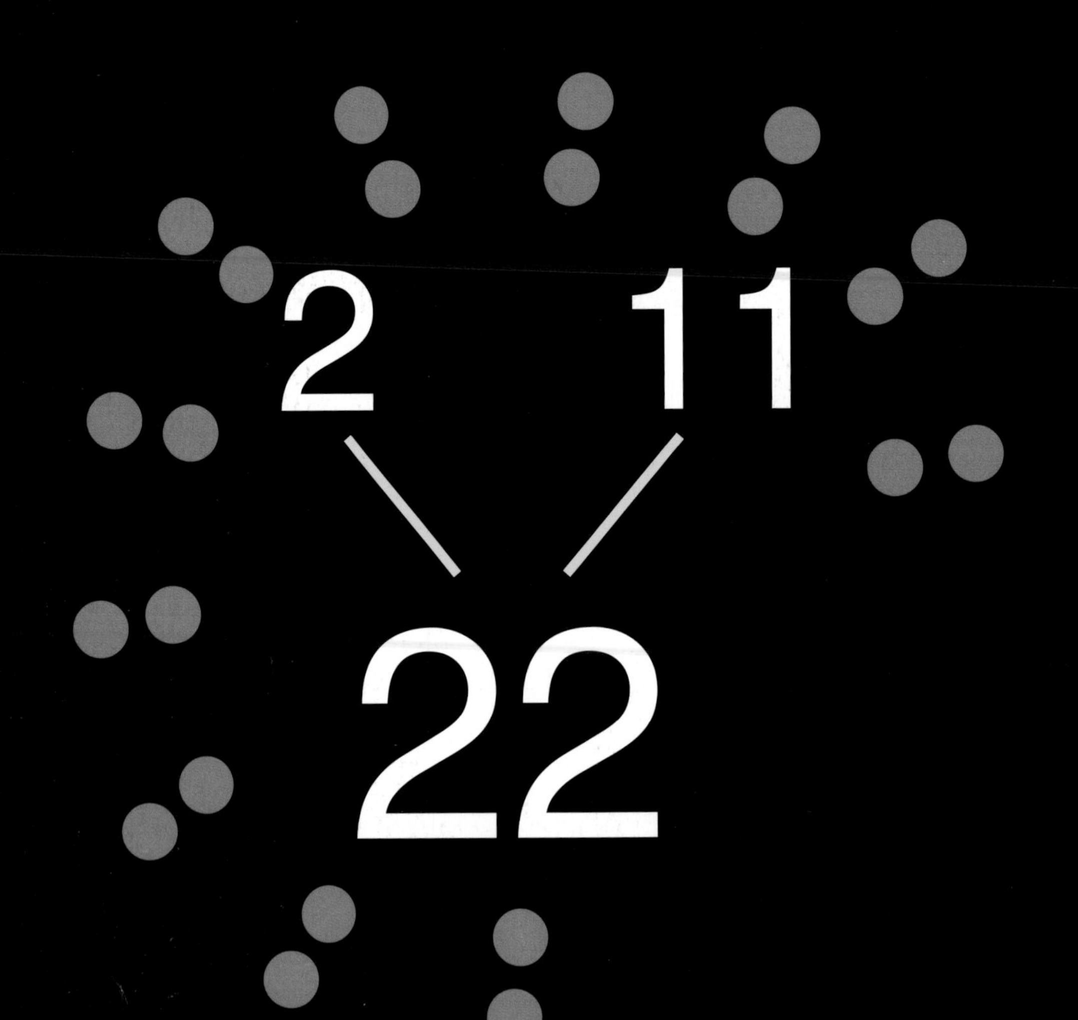
2
11
22

23

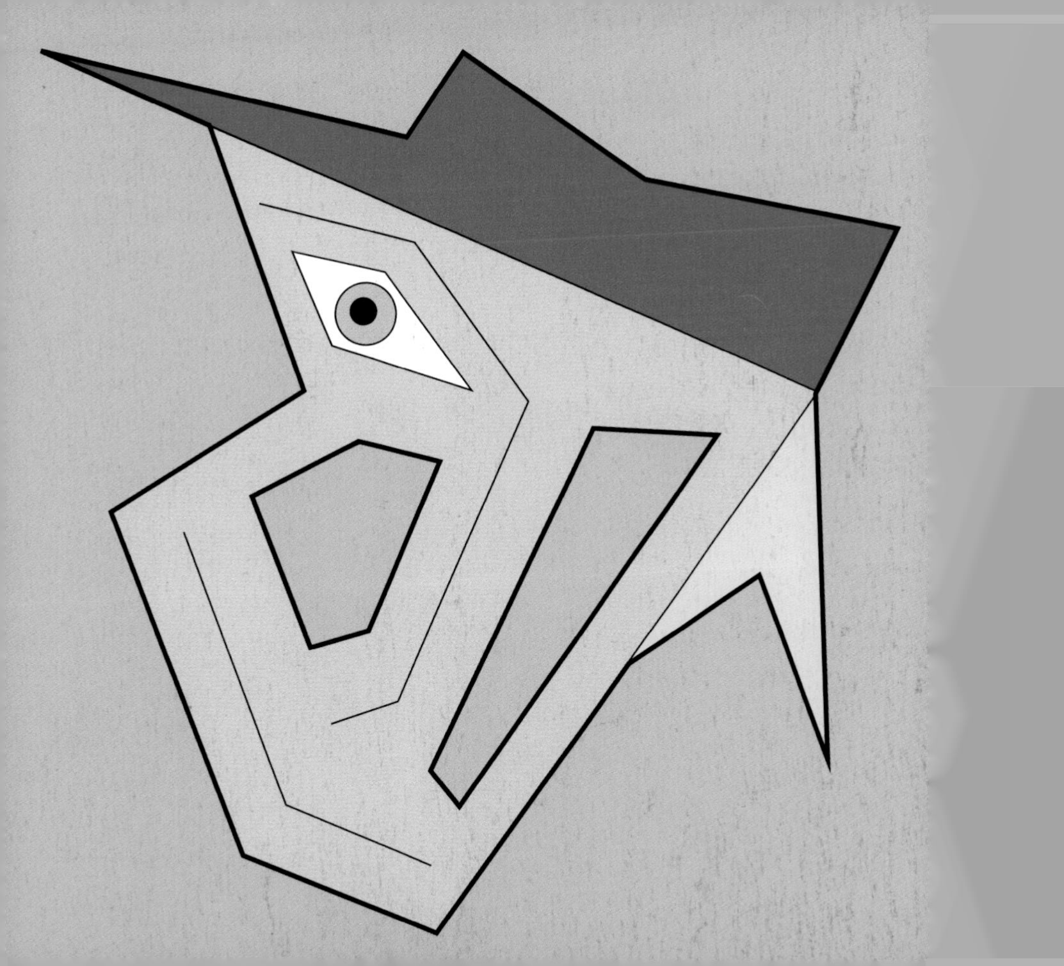

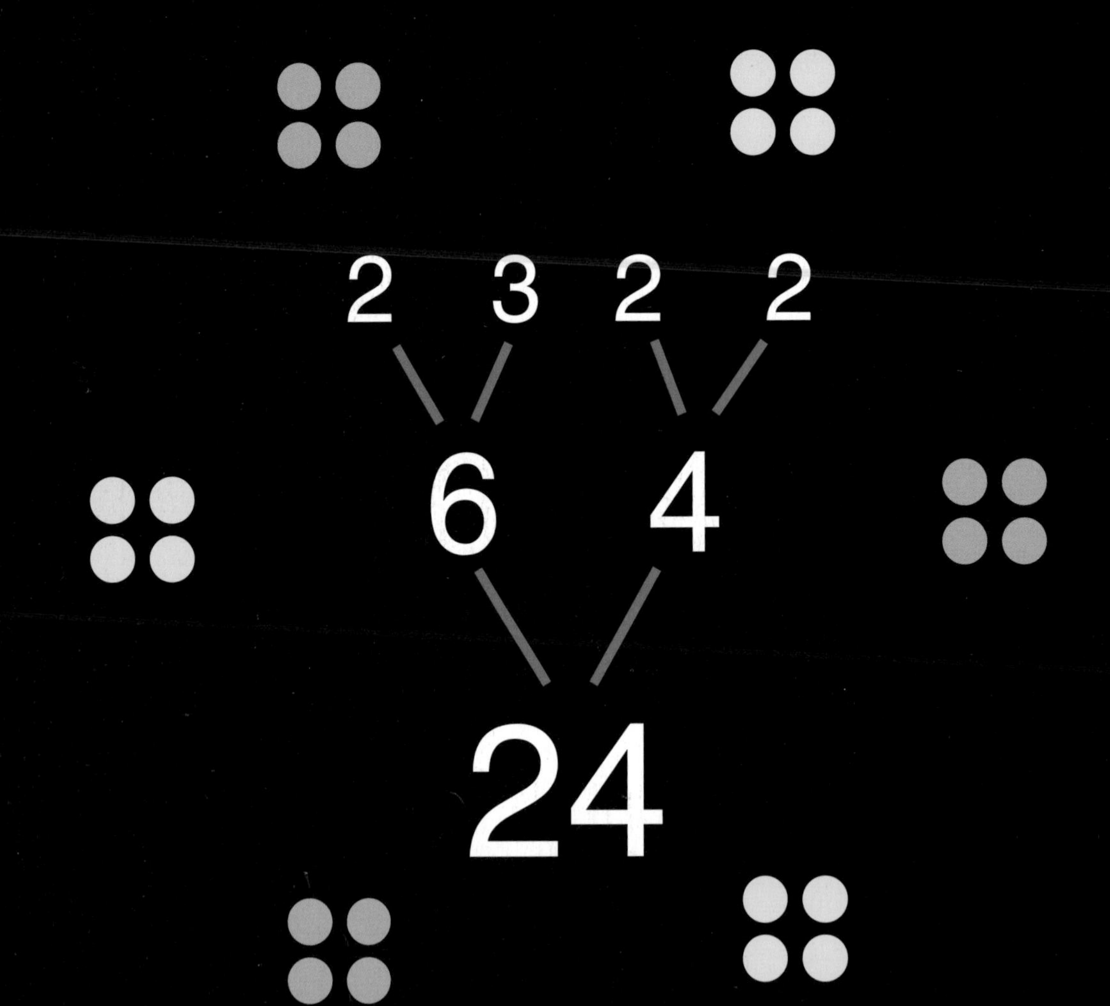
2
3
2
2
6
4
24

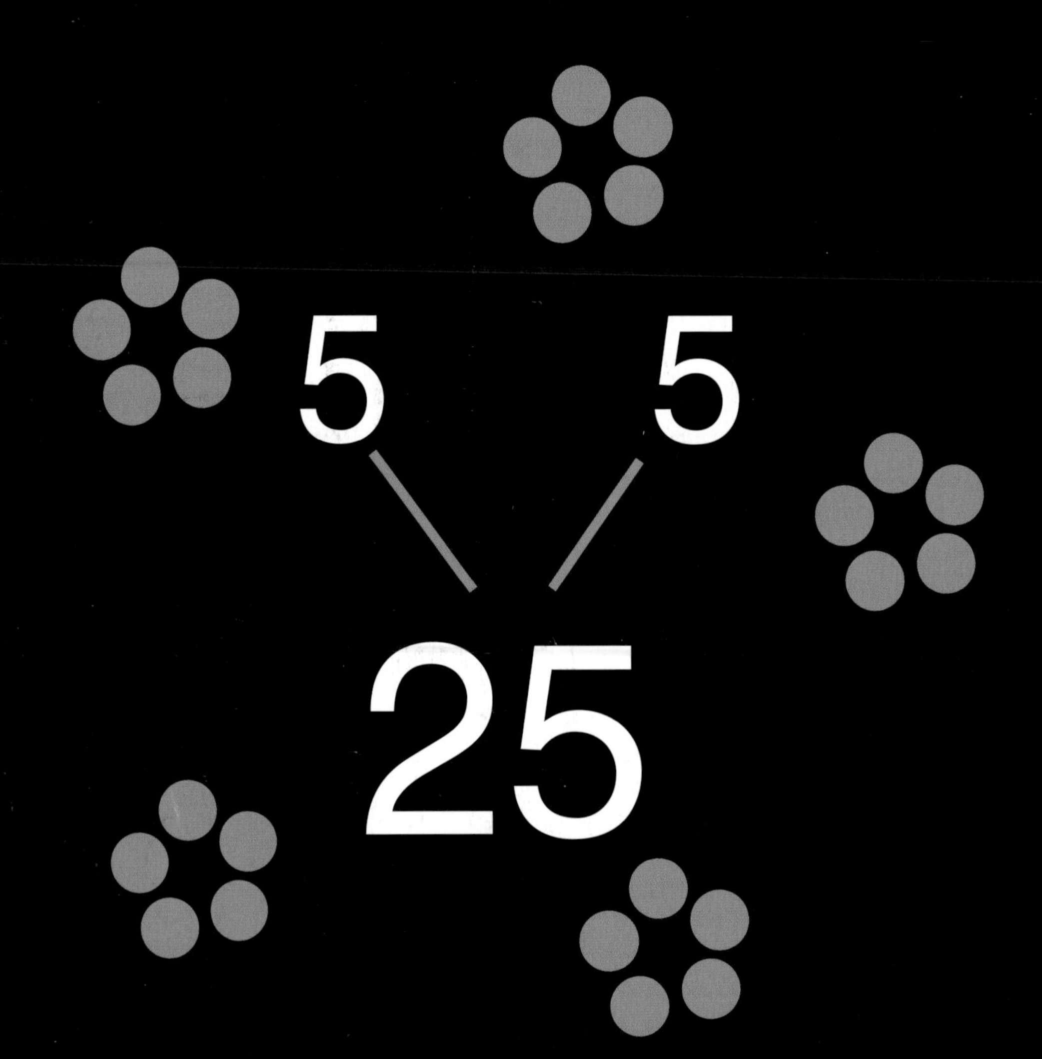
5
5
25

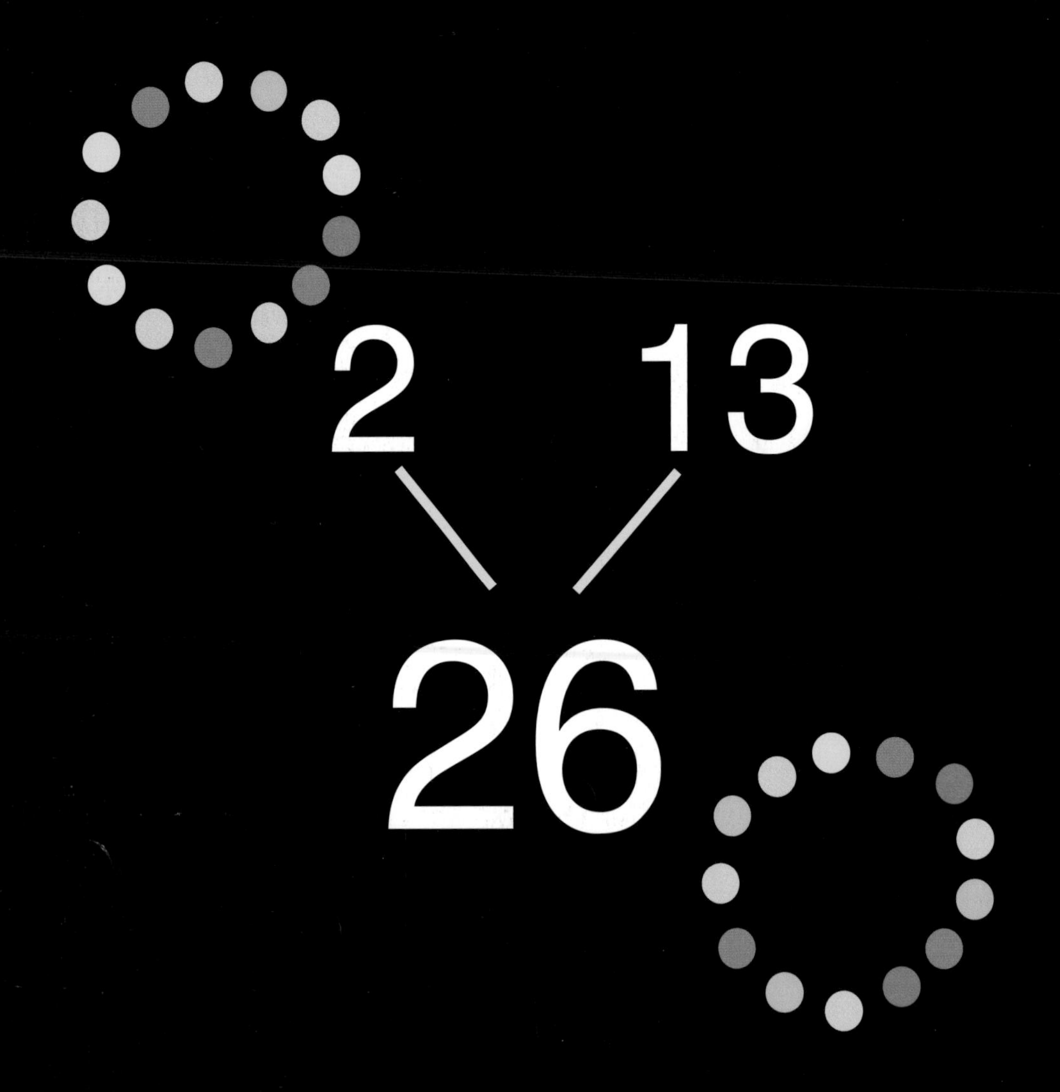
2
13
26

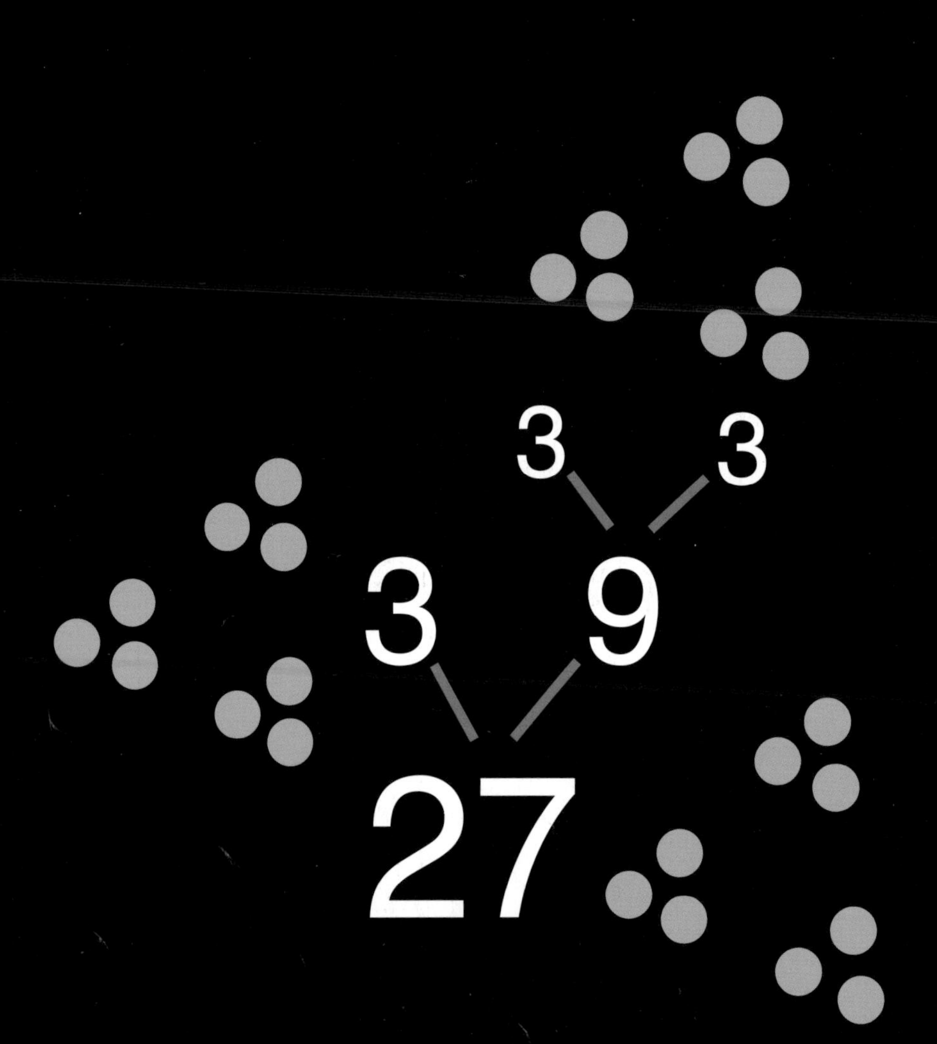
3
3
3
9
27

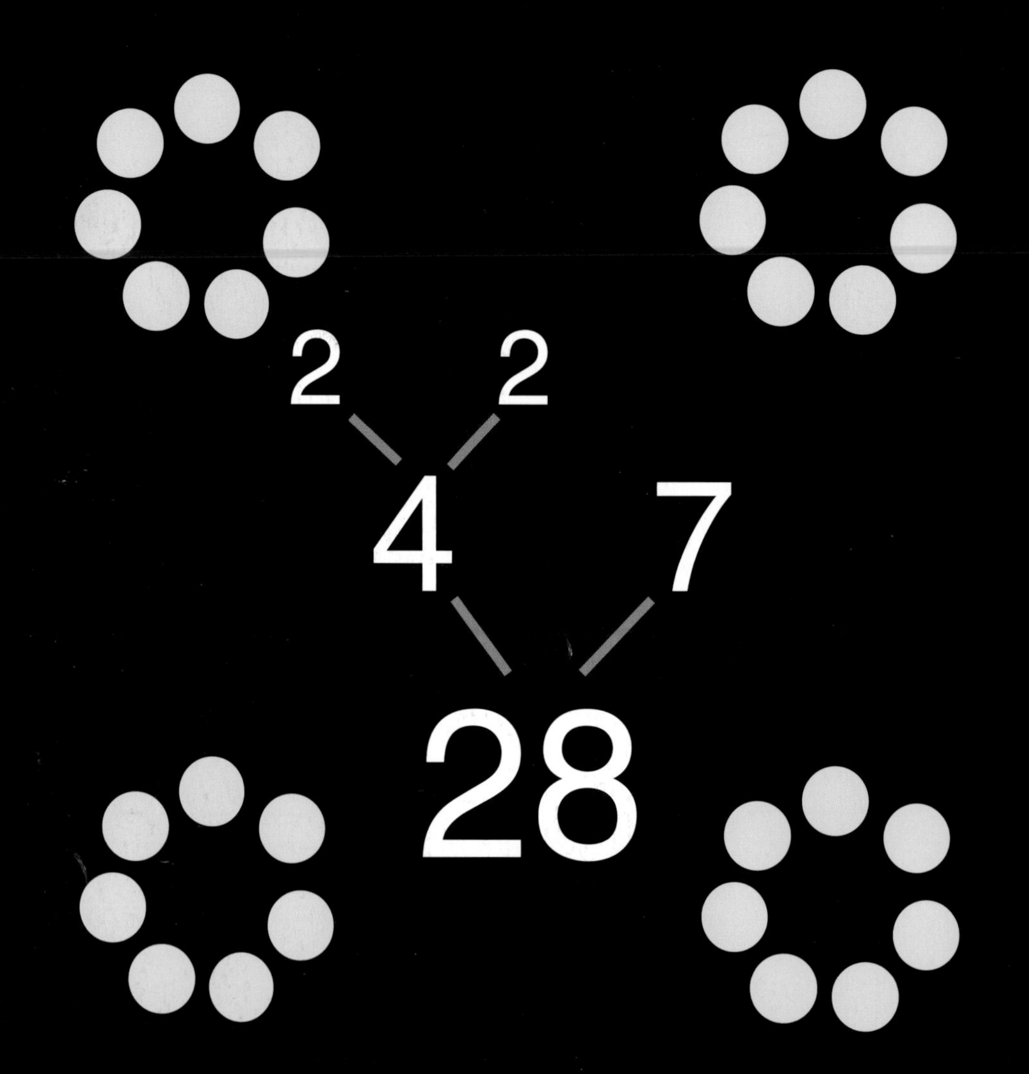
2
2
4
7
28

29

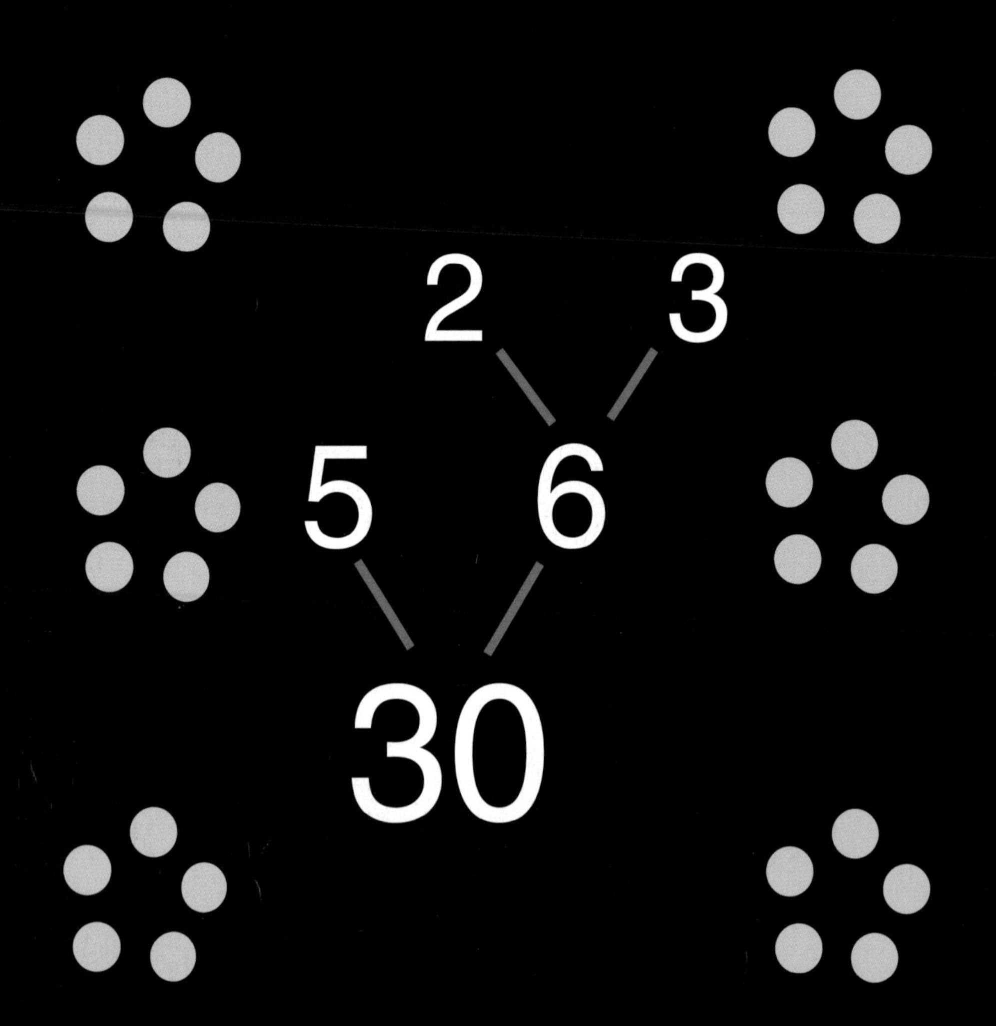
2
3
5
6
30

31

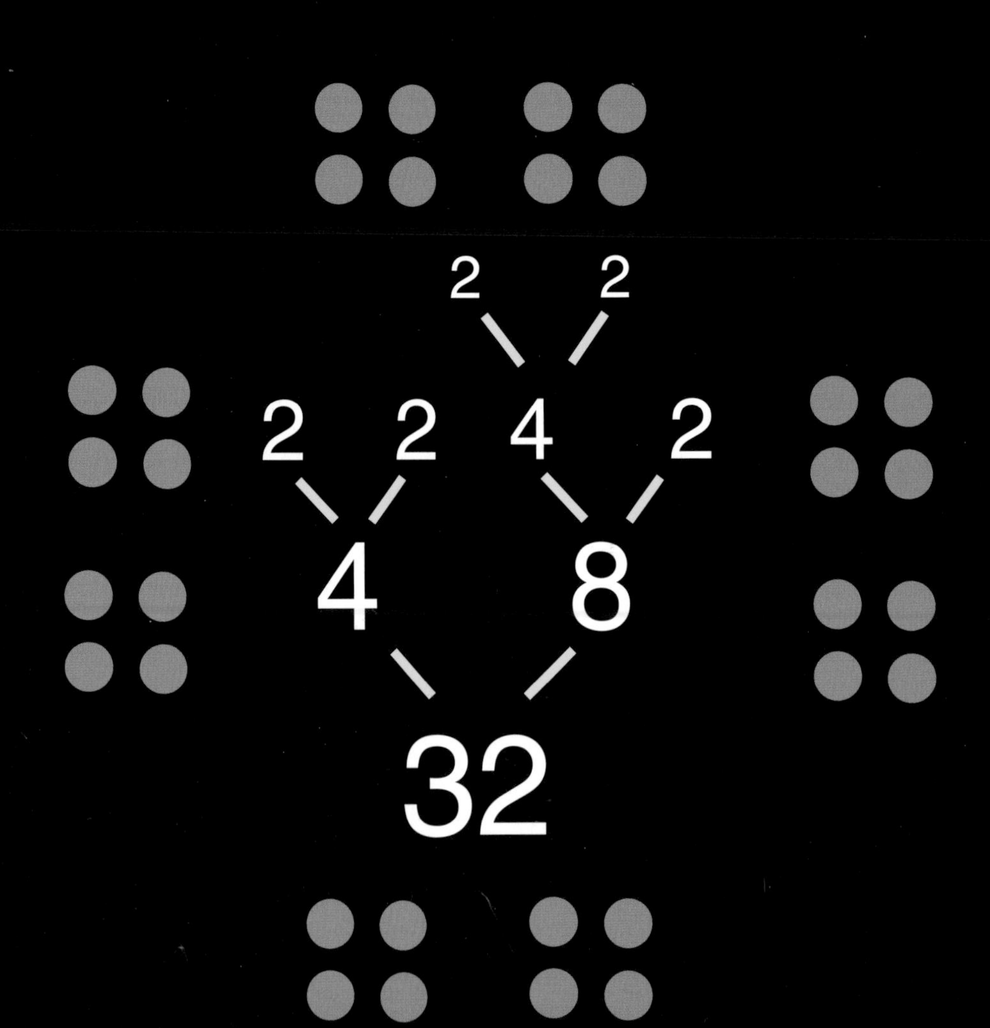
2 2
2 2 4 2
4 8
32

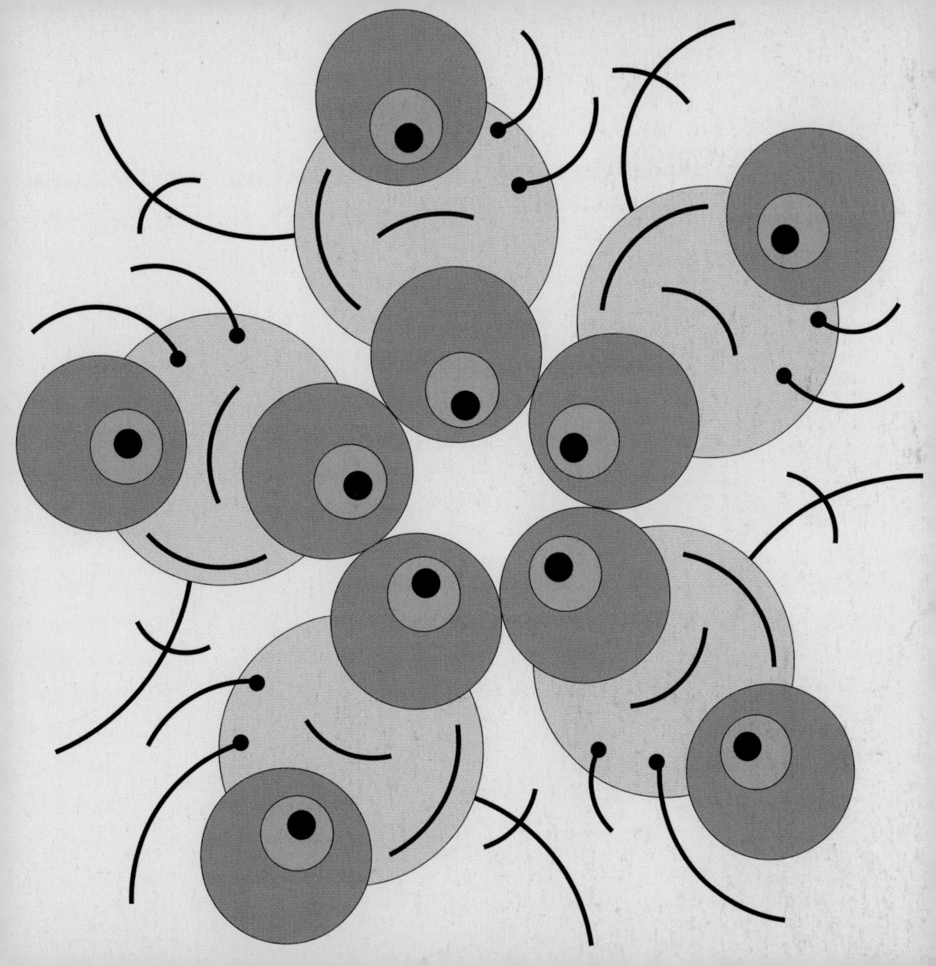

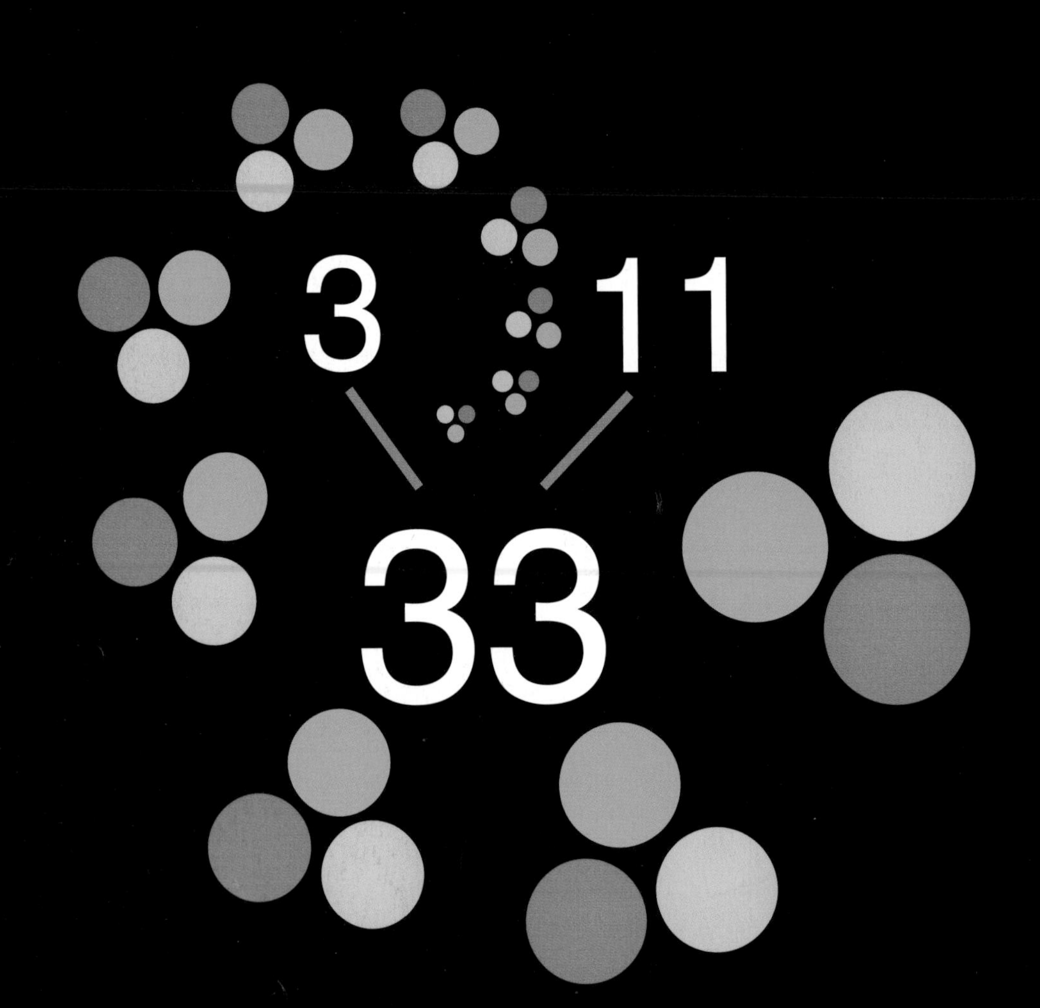
3
11
33

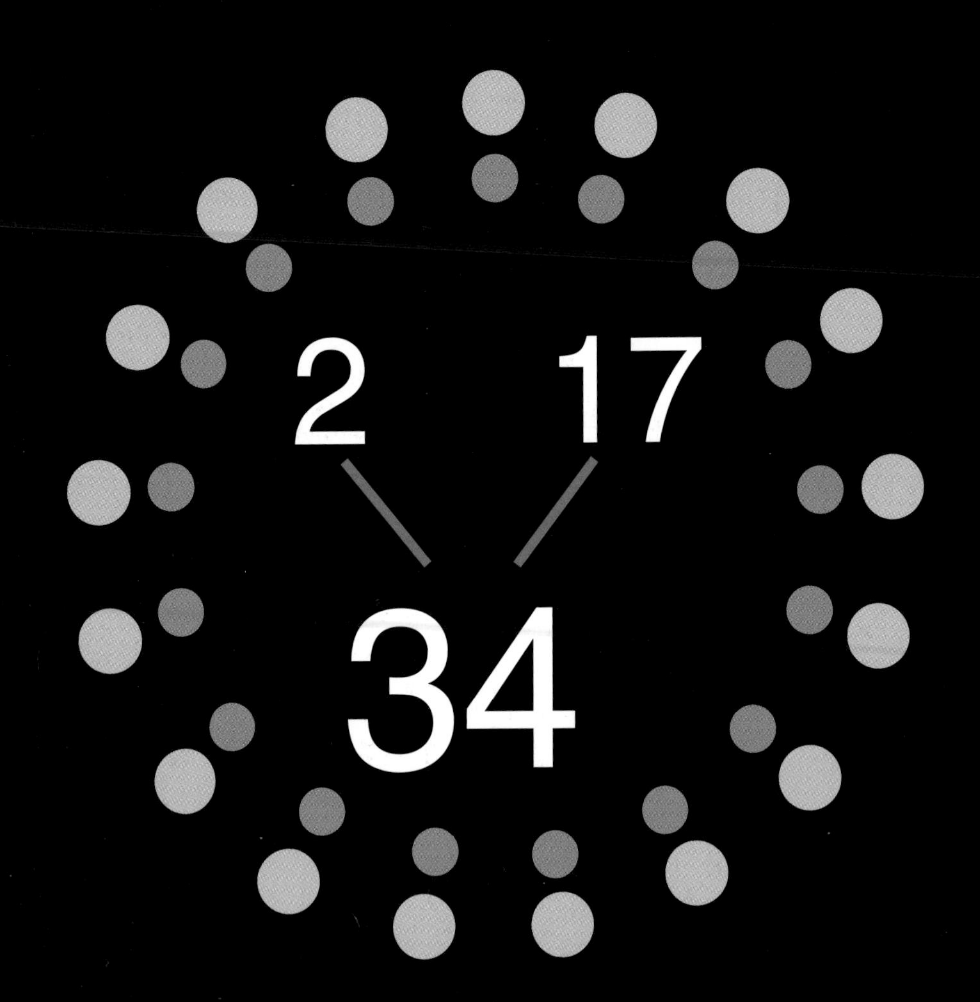
2
17
34

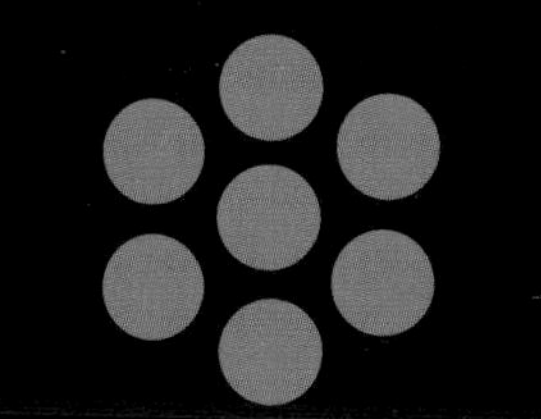

5
7

35

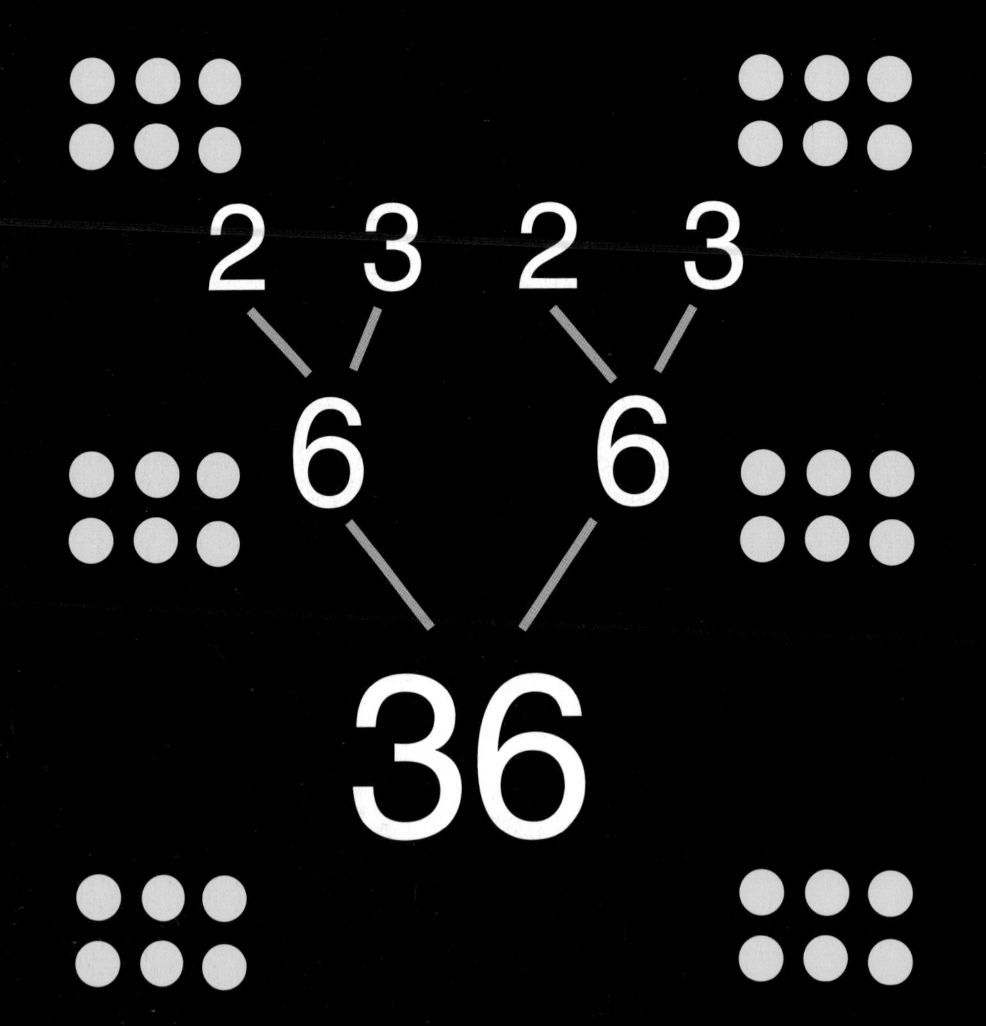
2 3 2 3
6 6
36

37

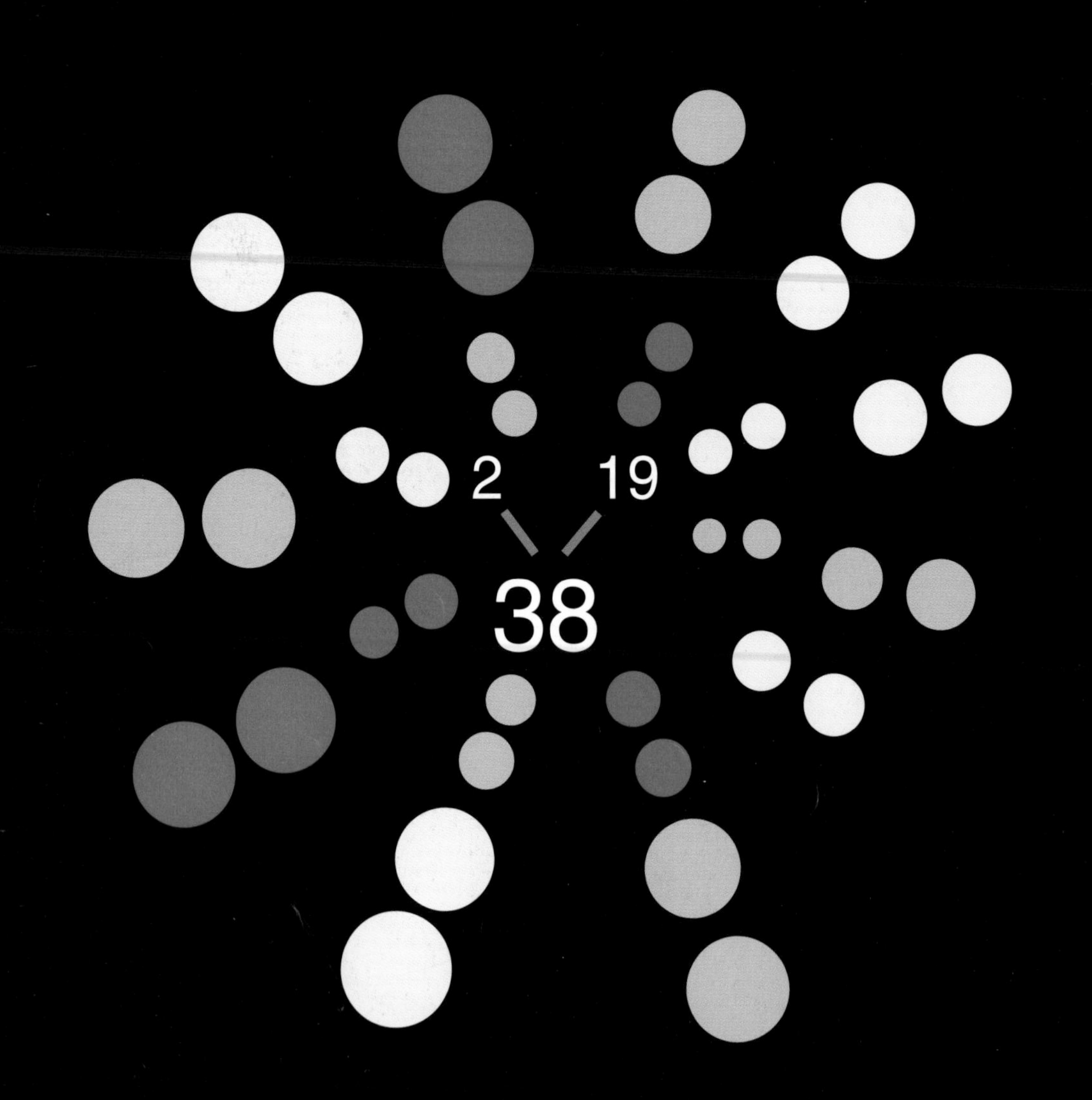
2
19
38

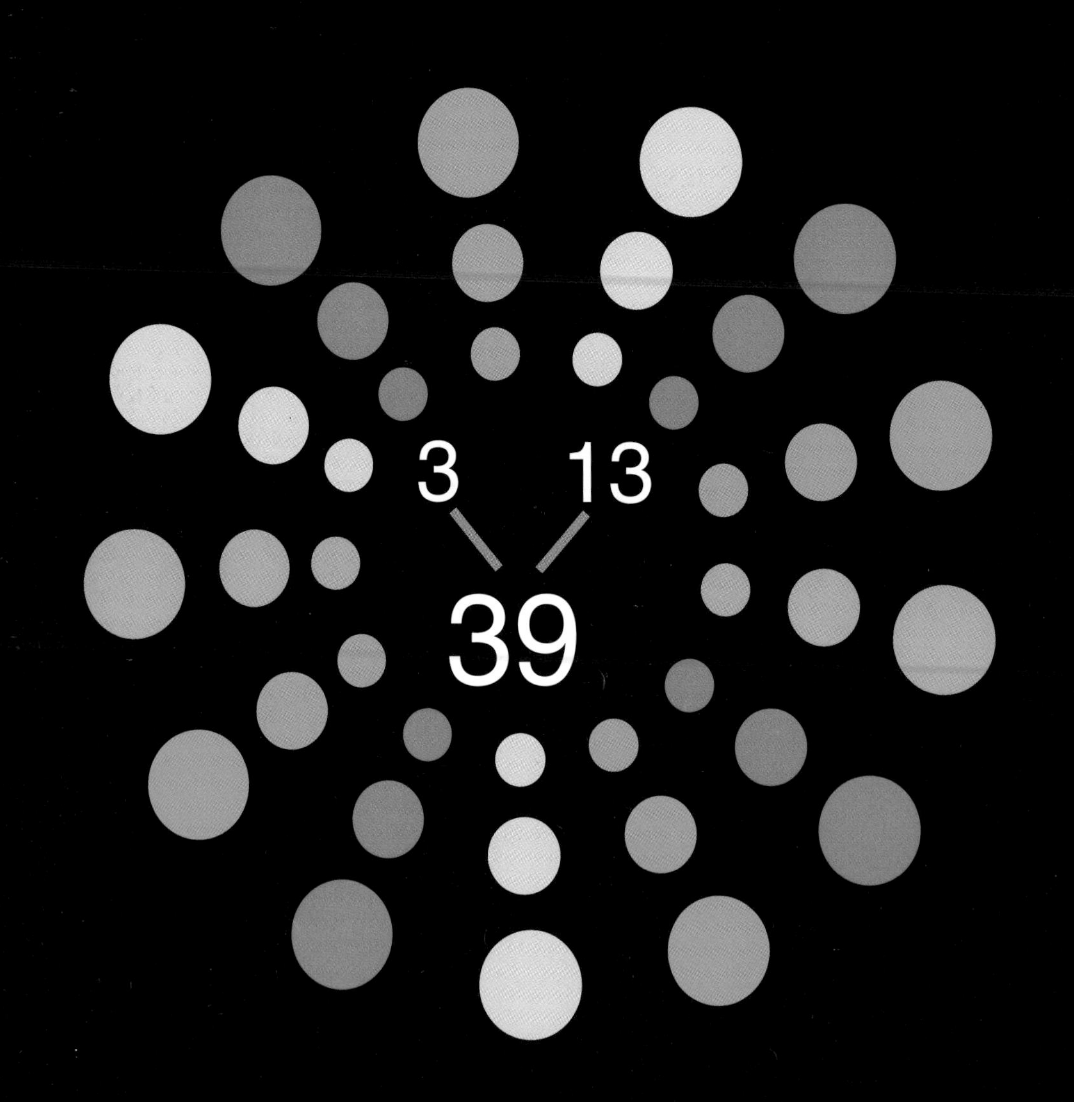
3
13
39

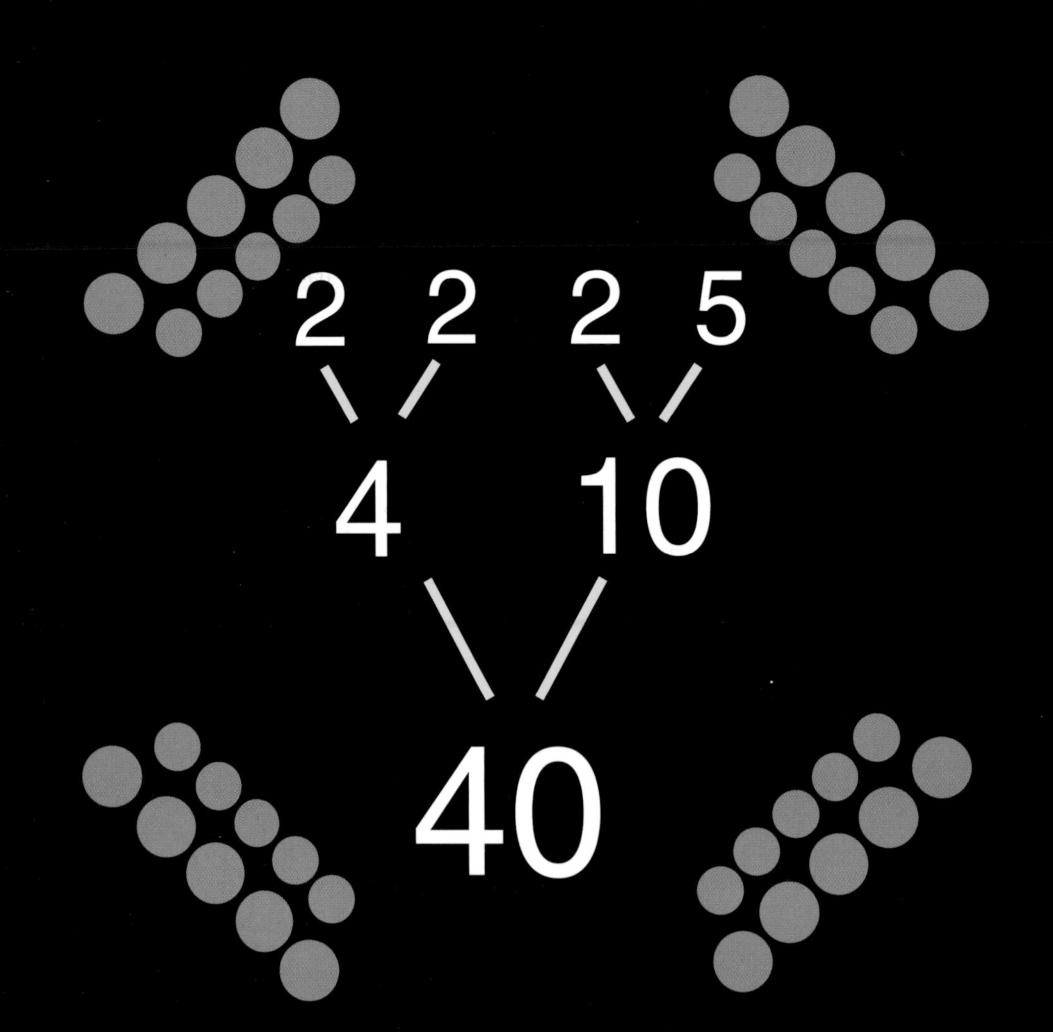
2 2 2 5
4 10
40

41

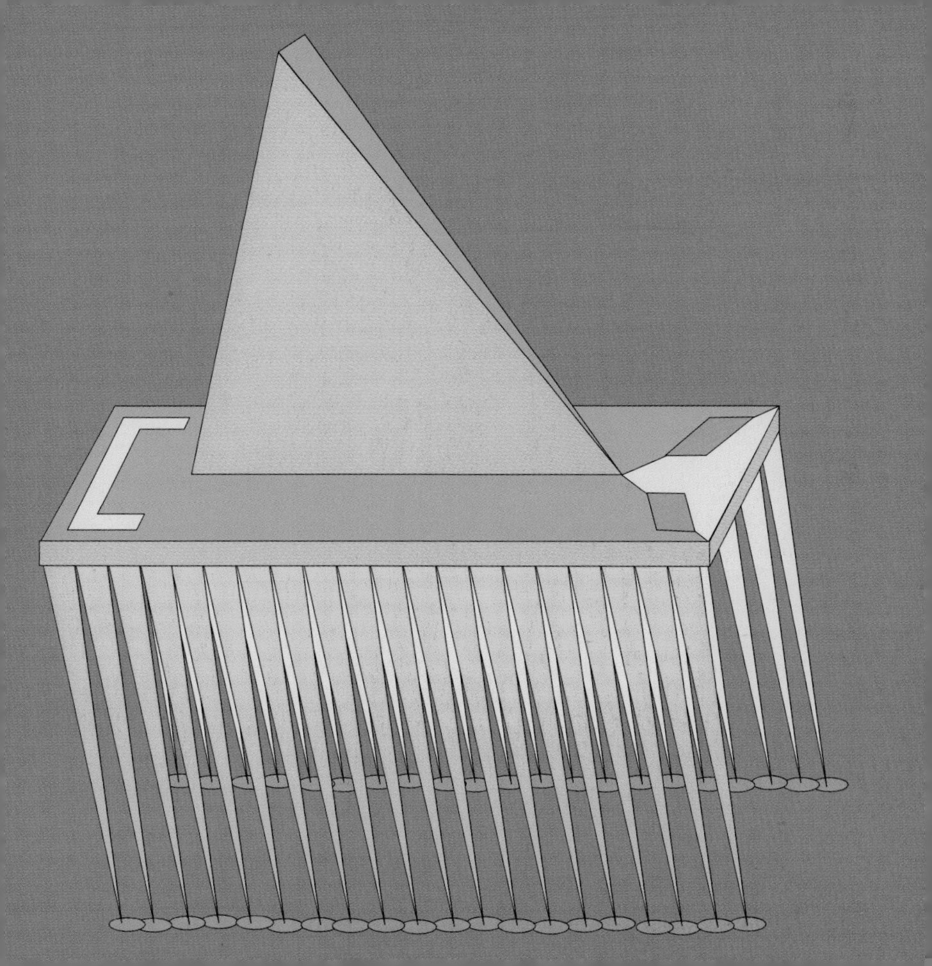

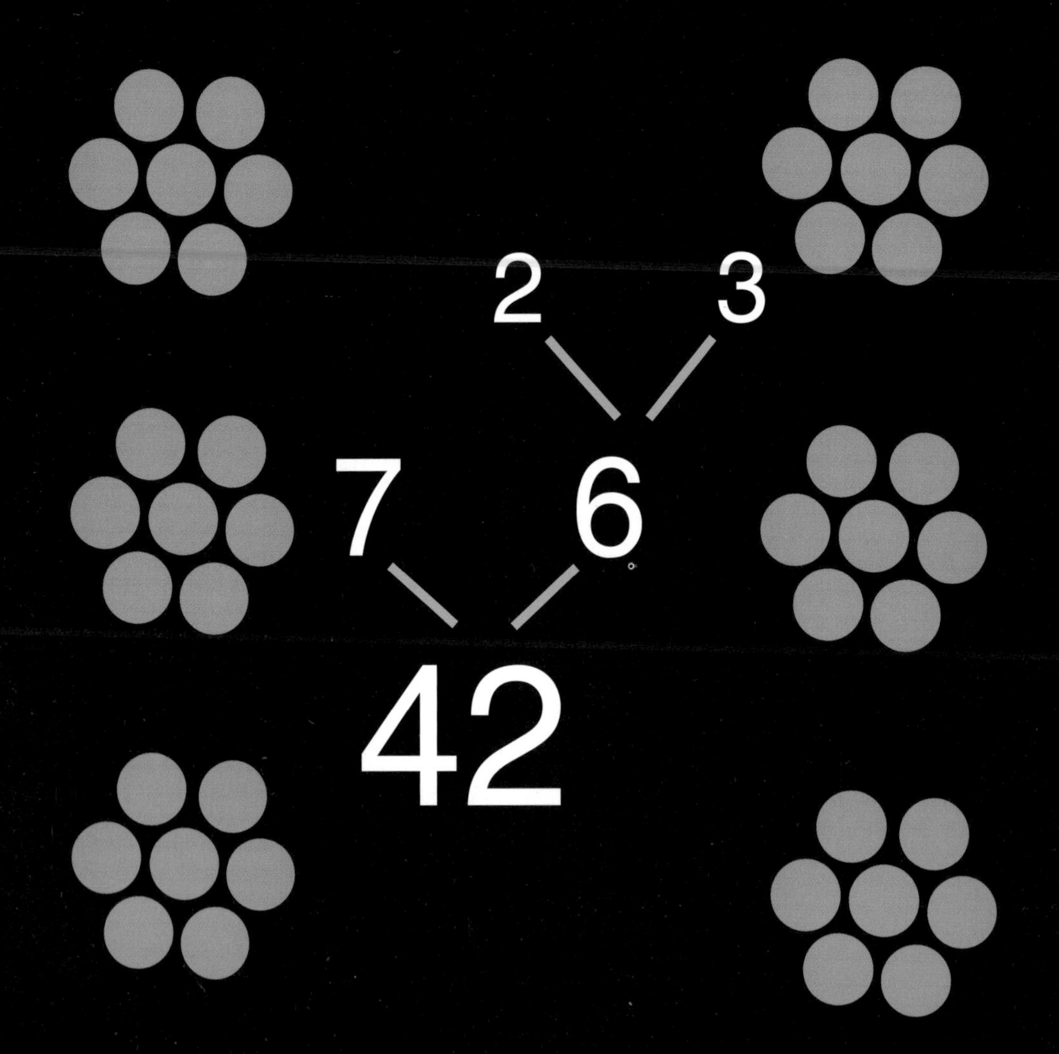
2
3
7
6
42

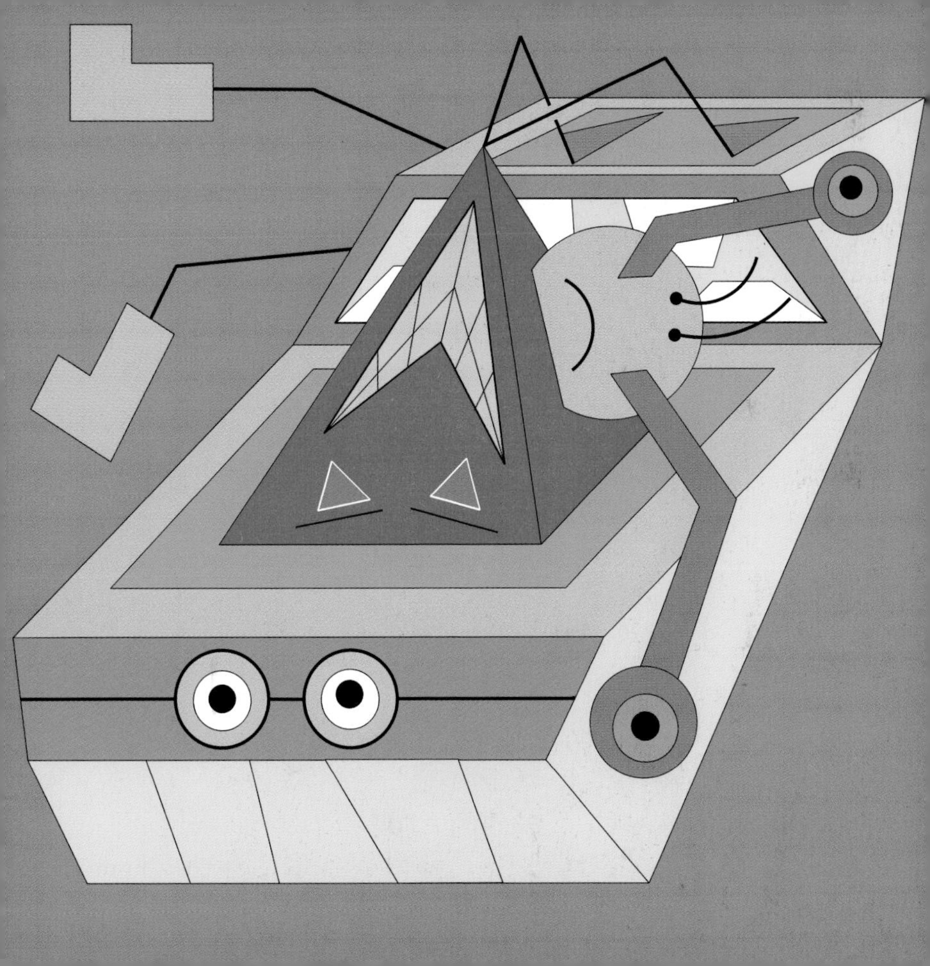

43

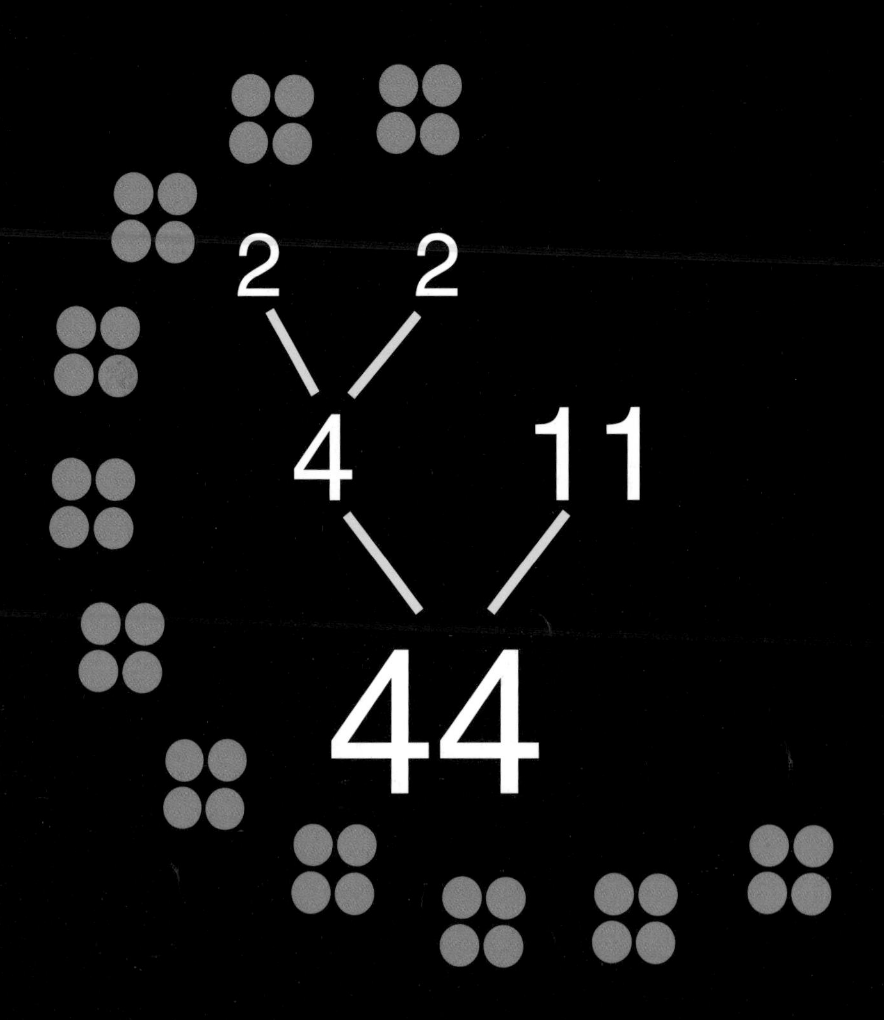
2
2
4
11
44

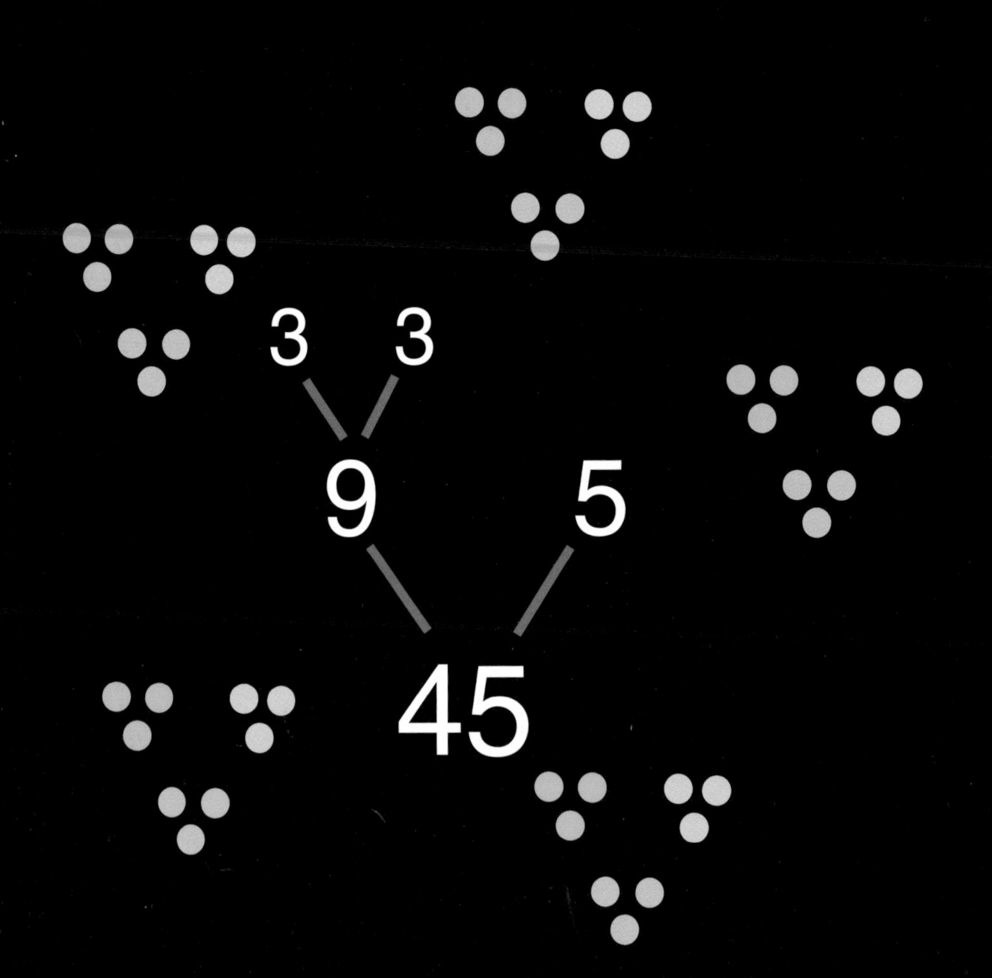
3
3
9
5
45

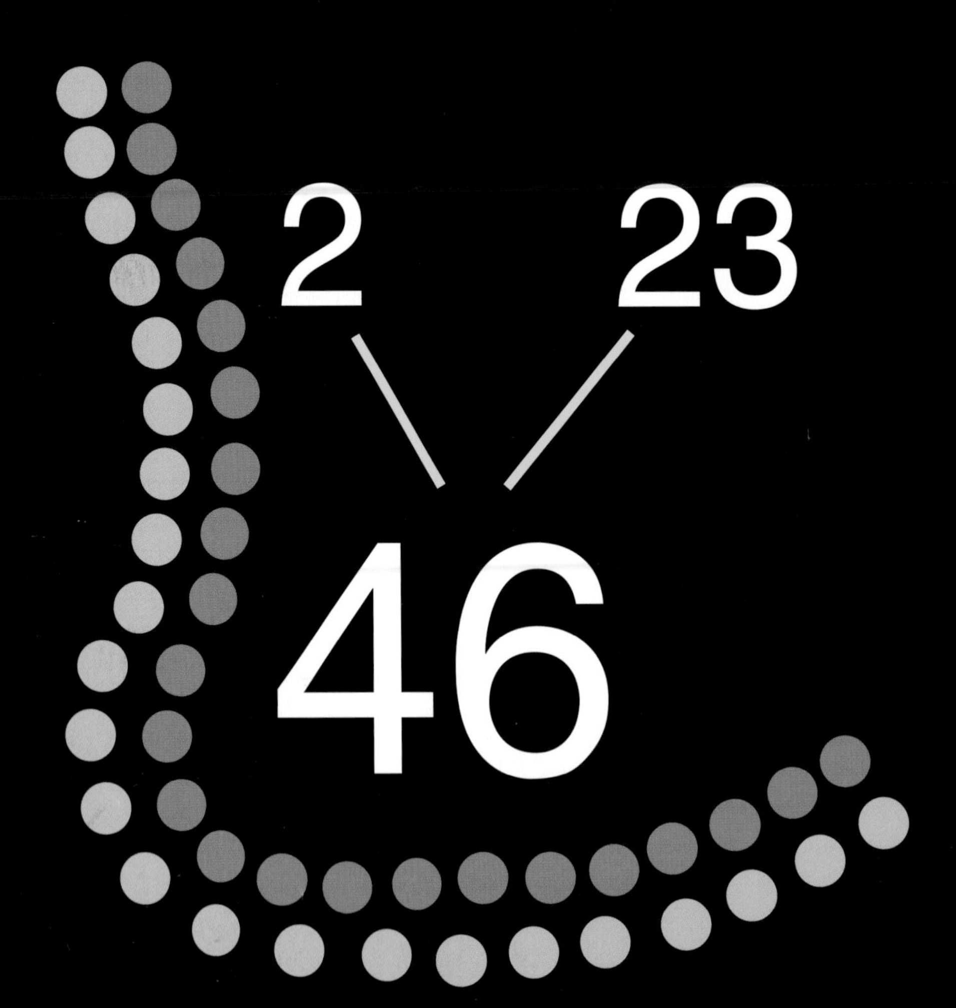
2
23
46

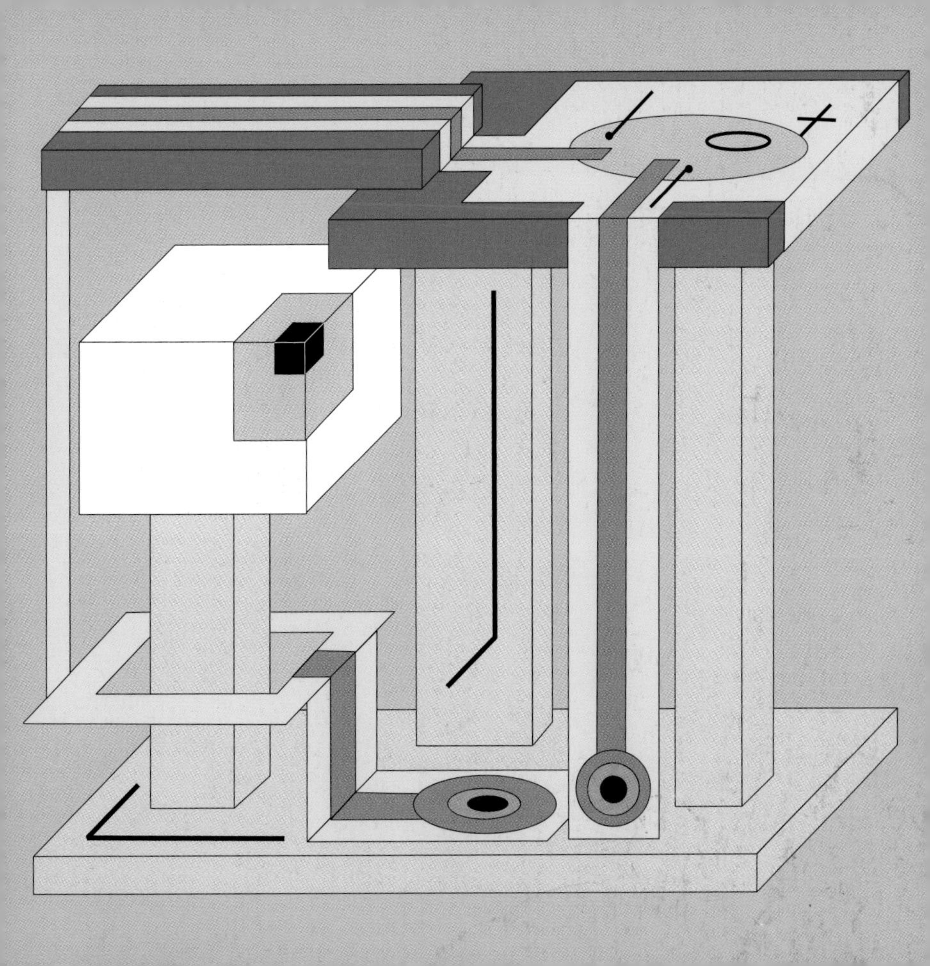

47

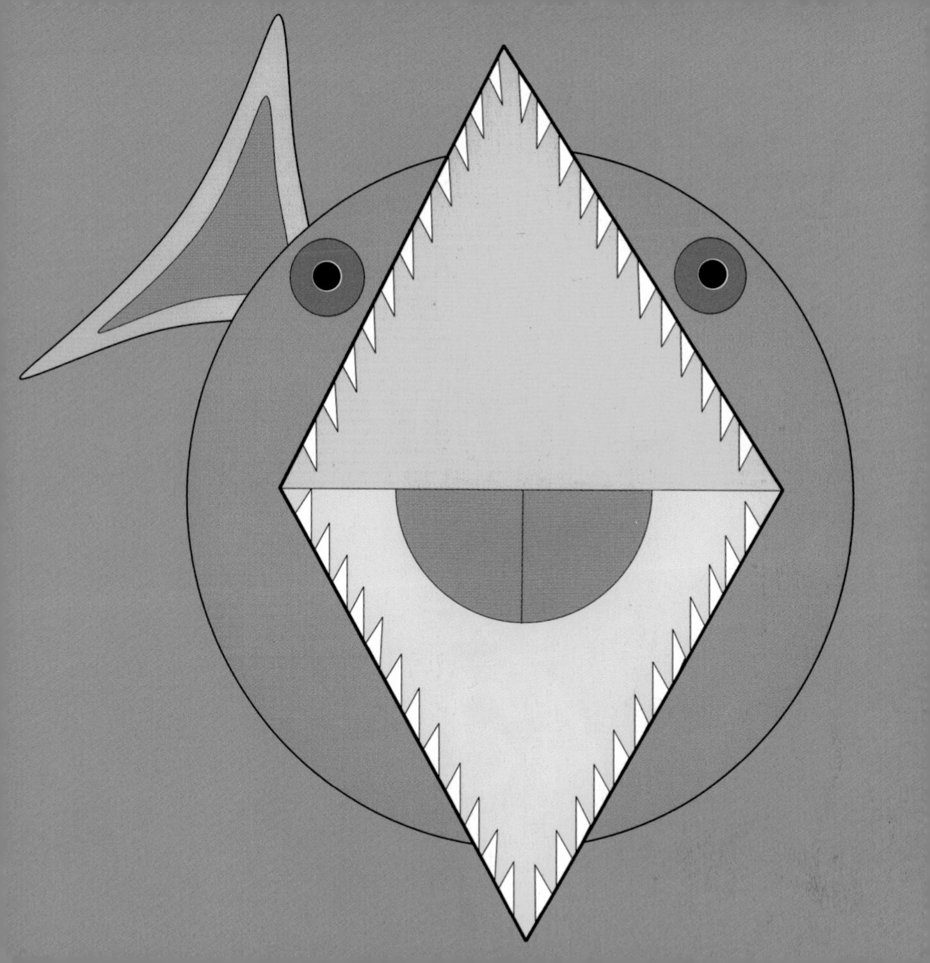

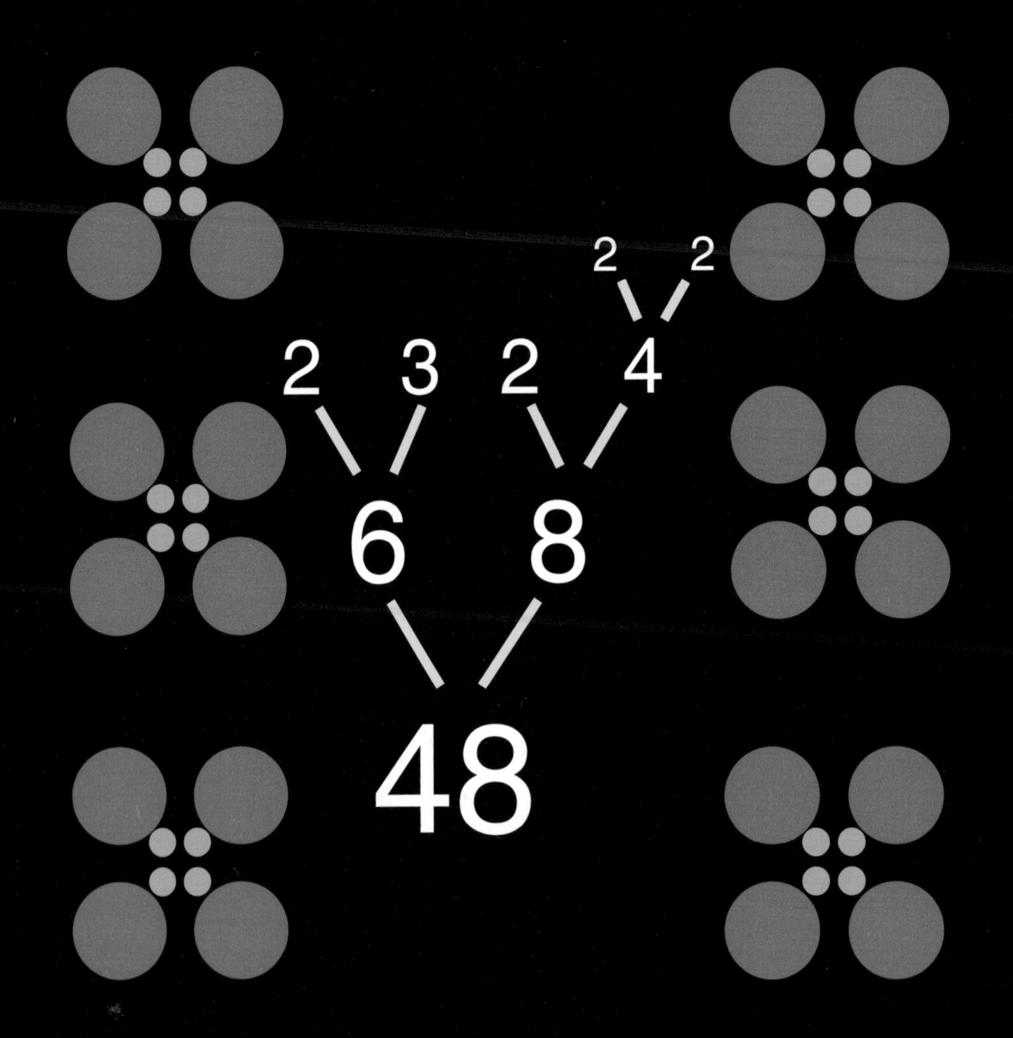
2
2
2
3
2
4
6
8
48

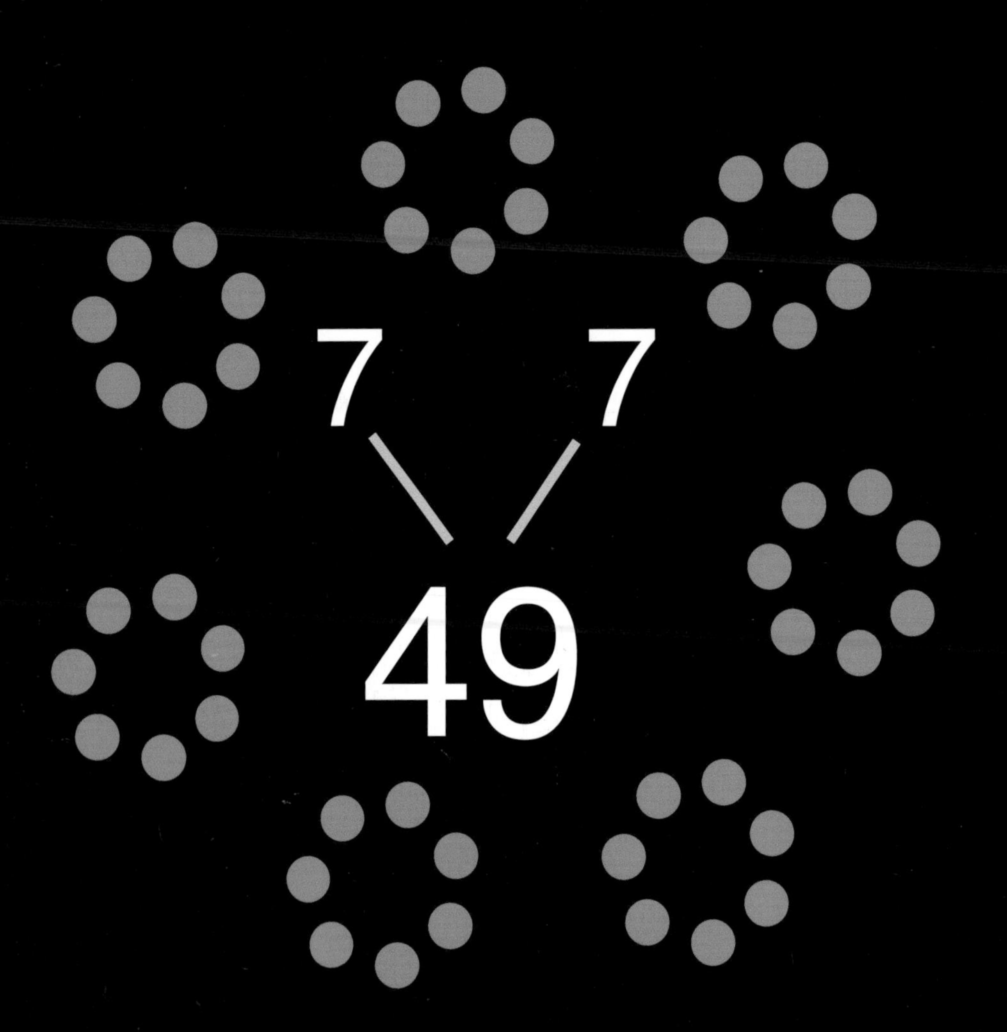
7
7
49

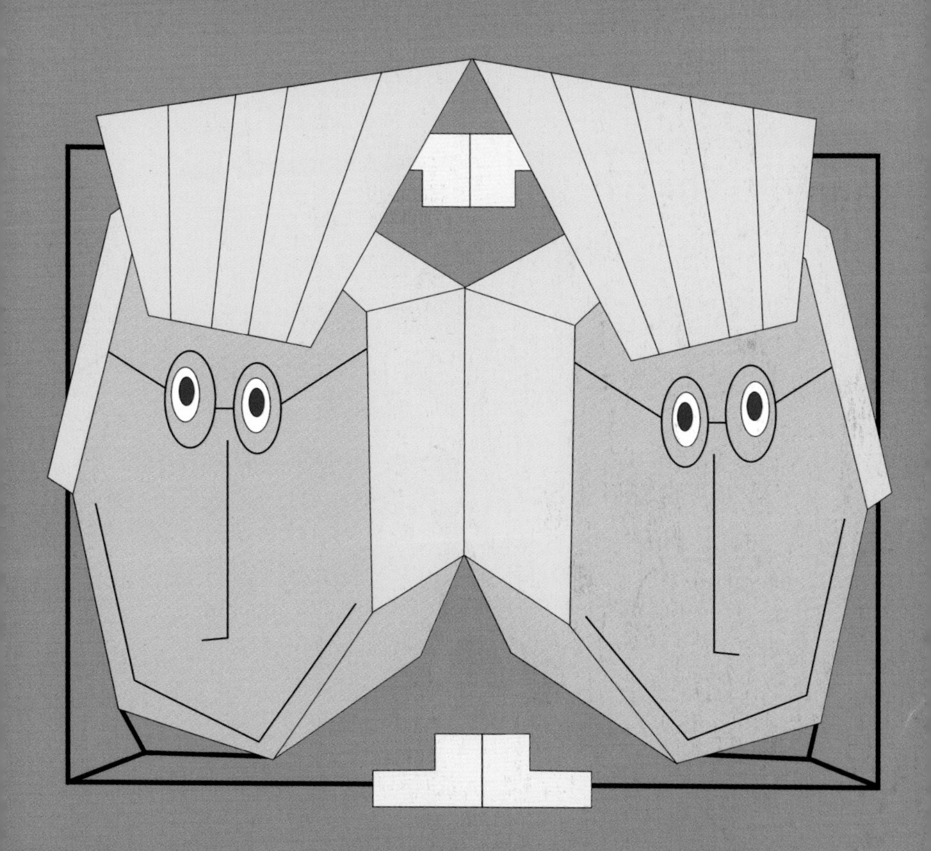

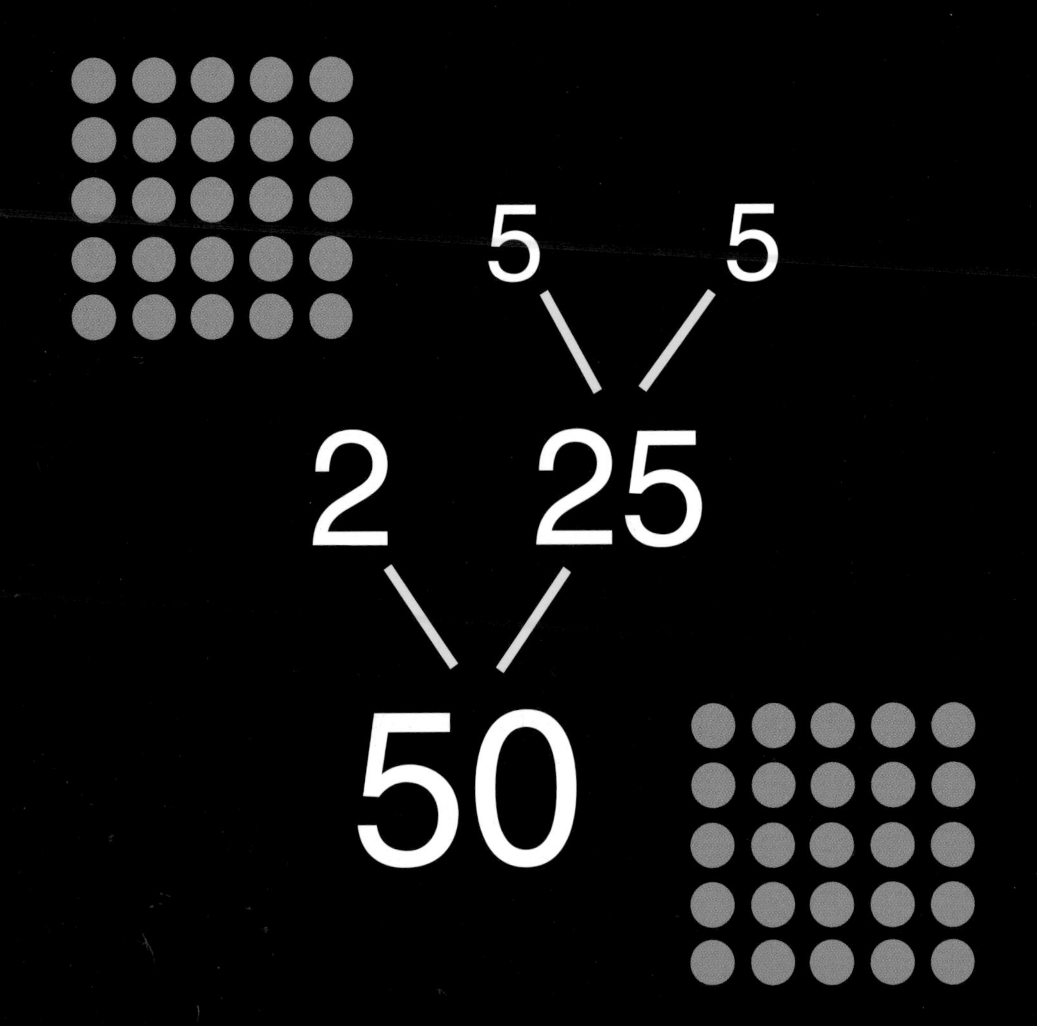
5
5
2
25
50

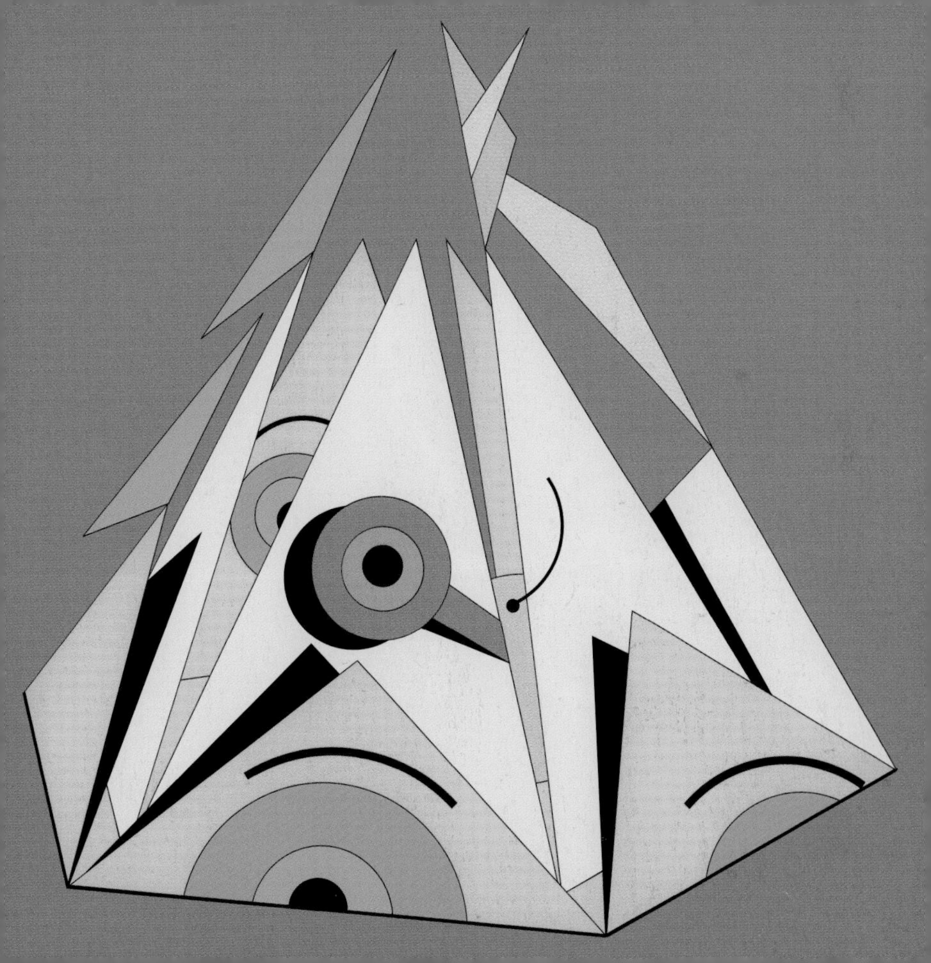

3
17
51

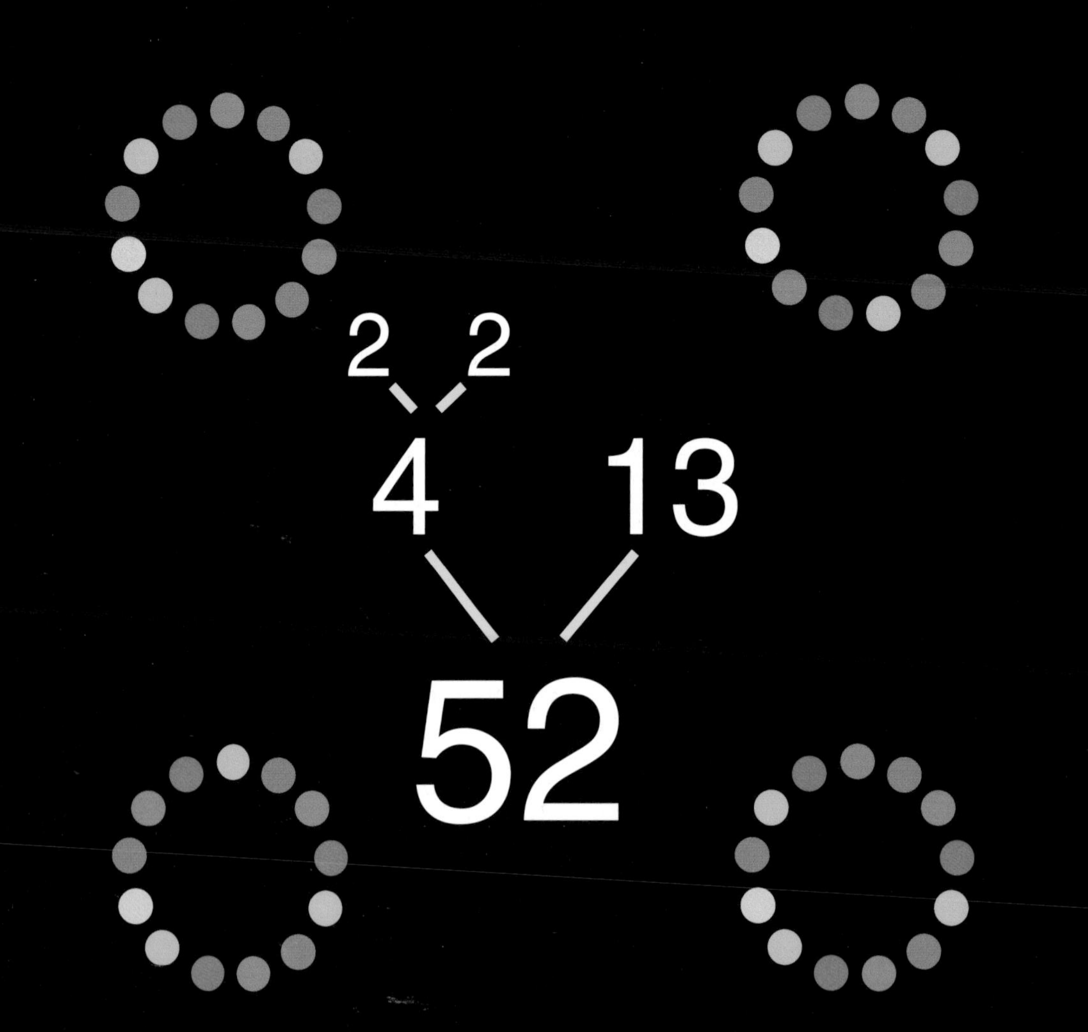
2
2
4
13
52

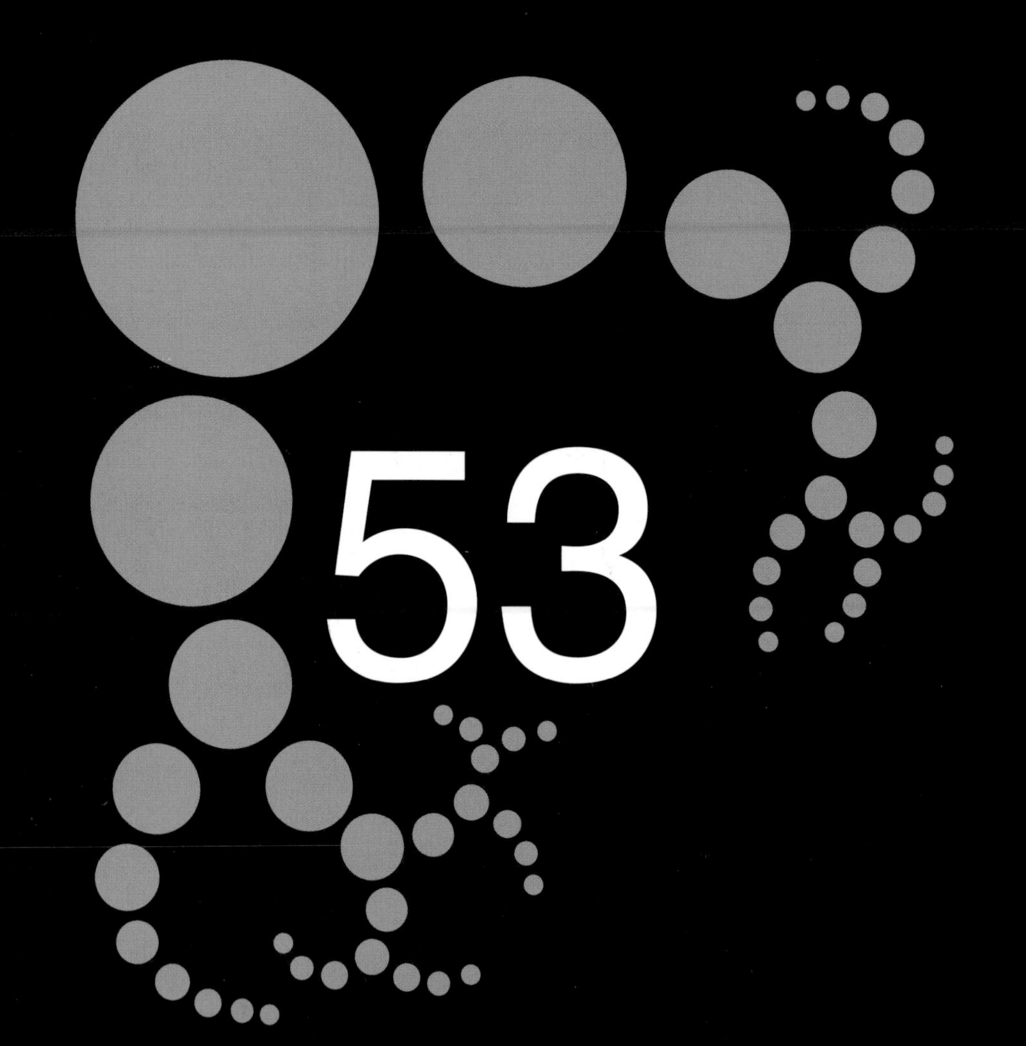
53

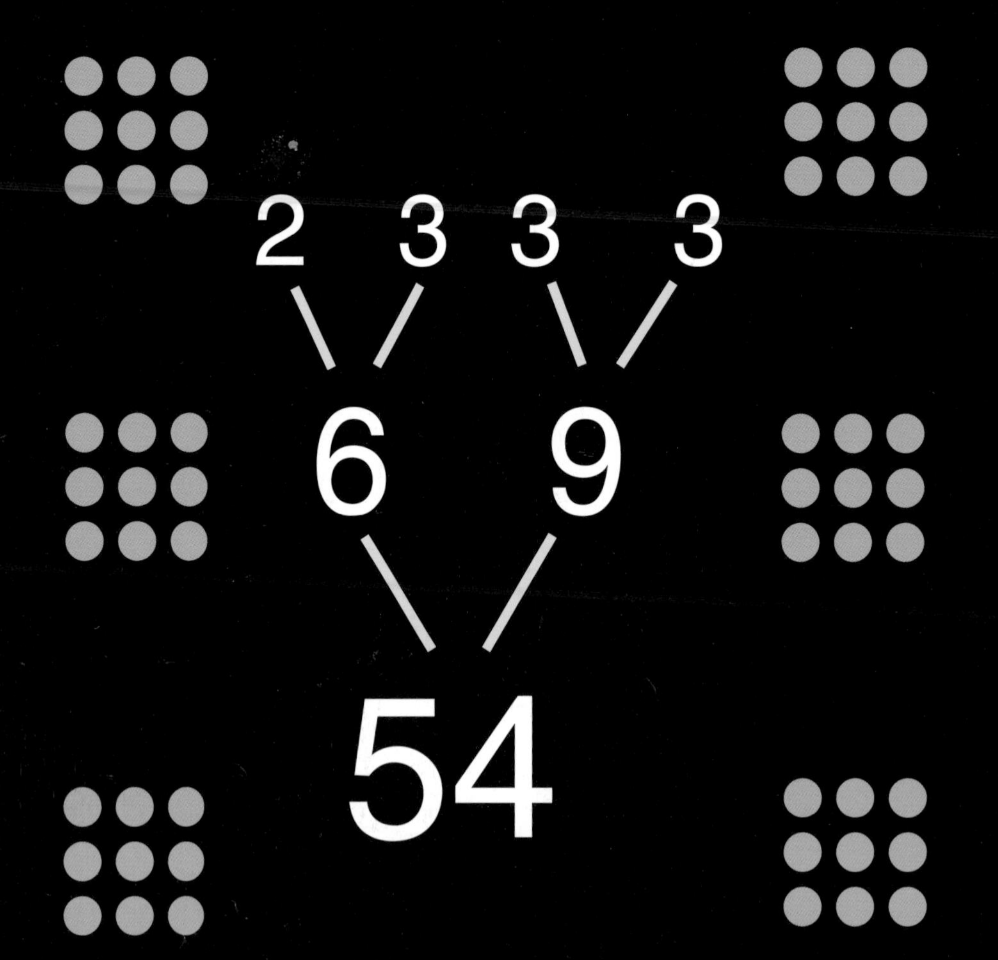
2 3 3 3
6 9
54

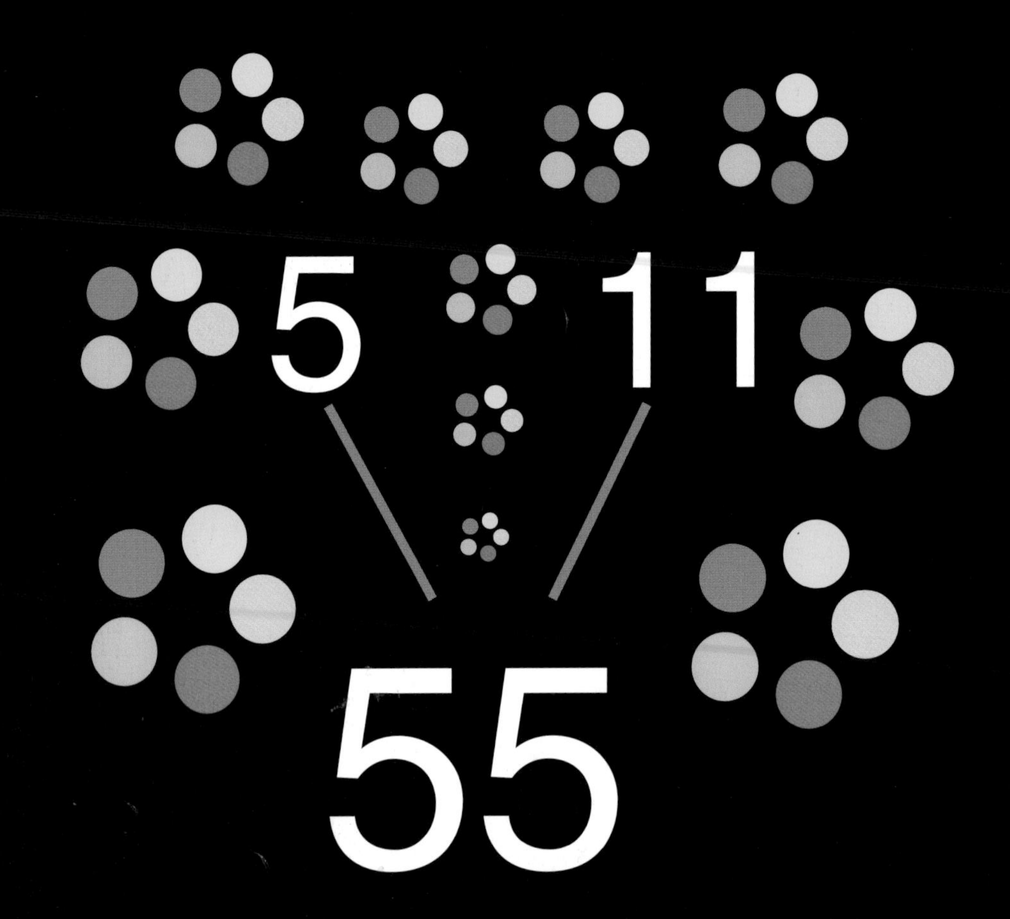
5
11
55

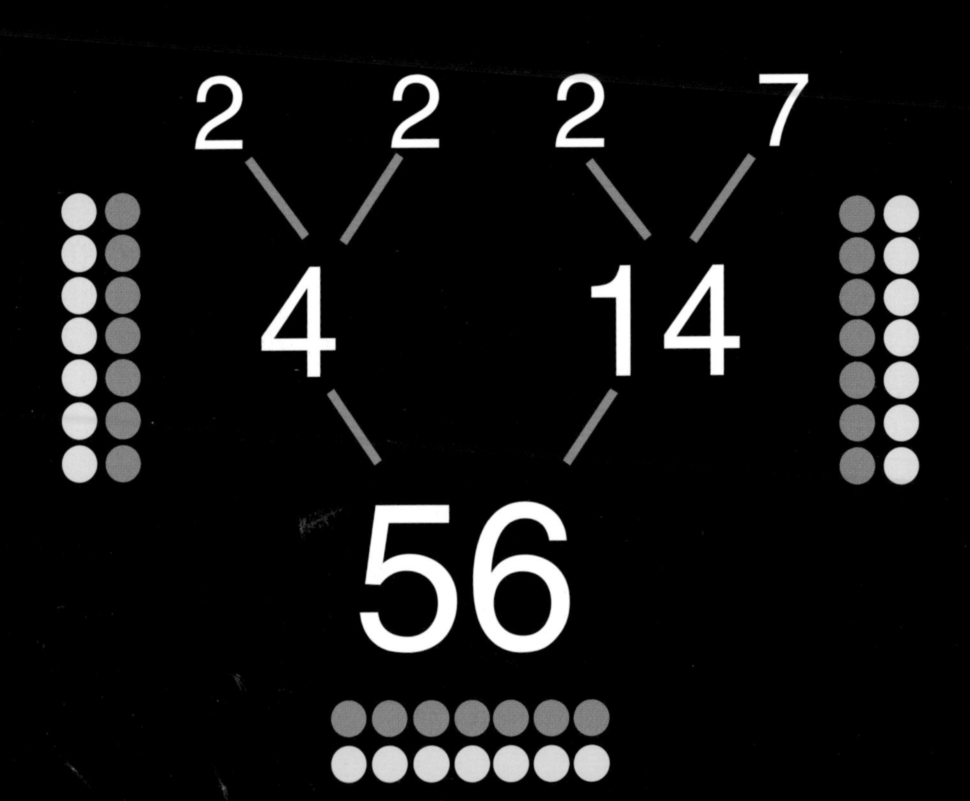
2
2
2
7
4
14
56

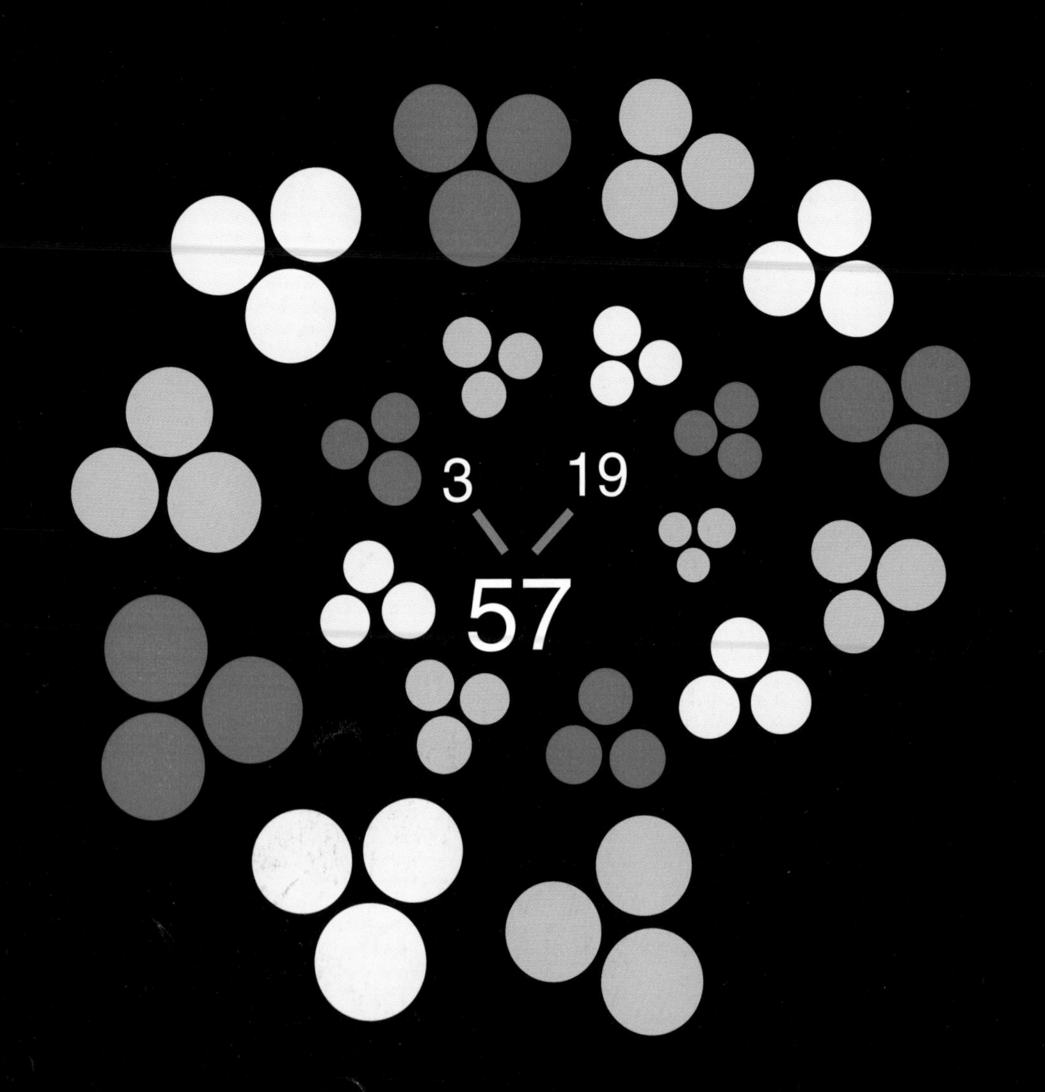
3
19
57

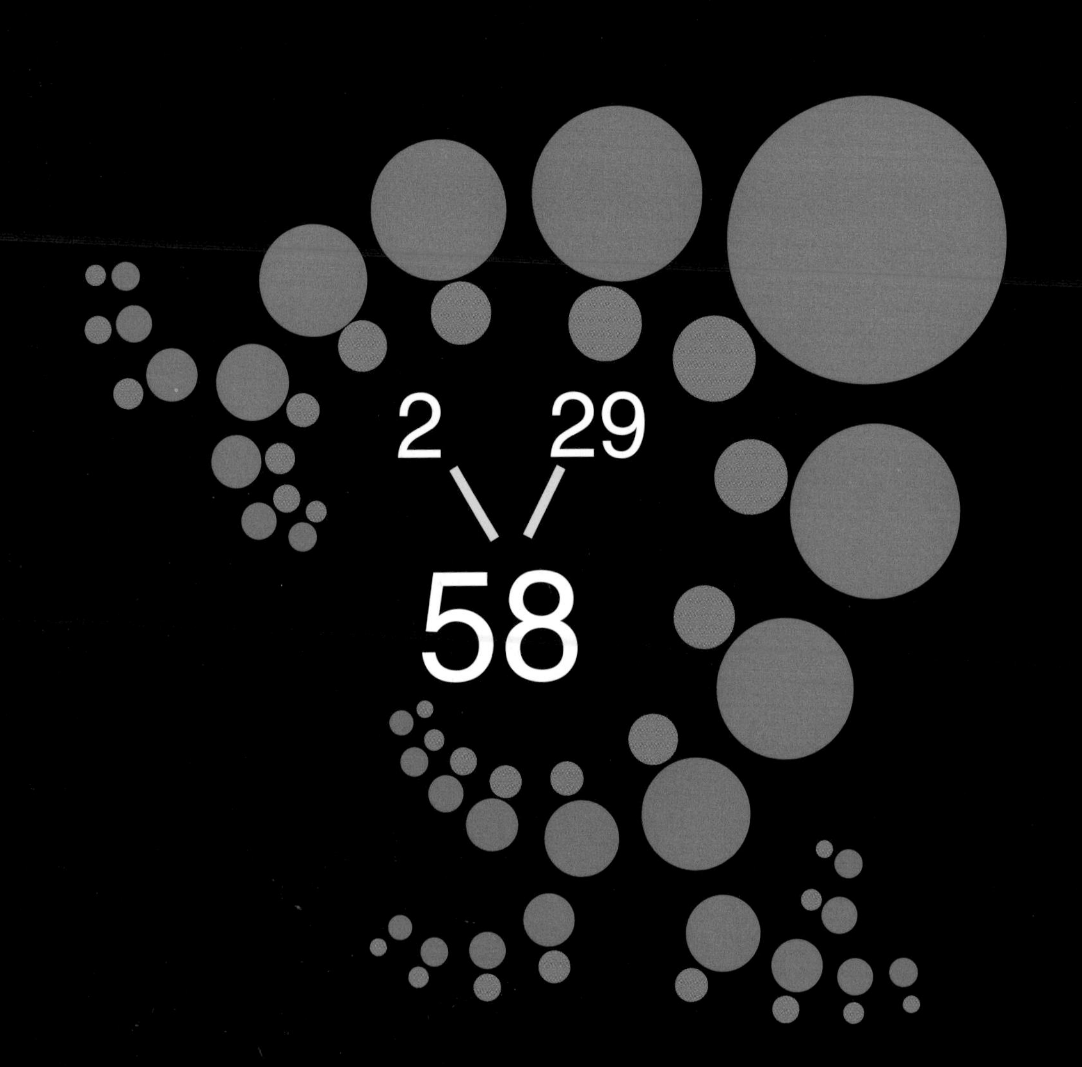
2
29
58

59

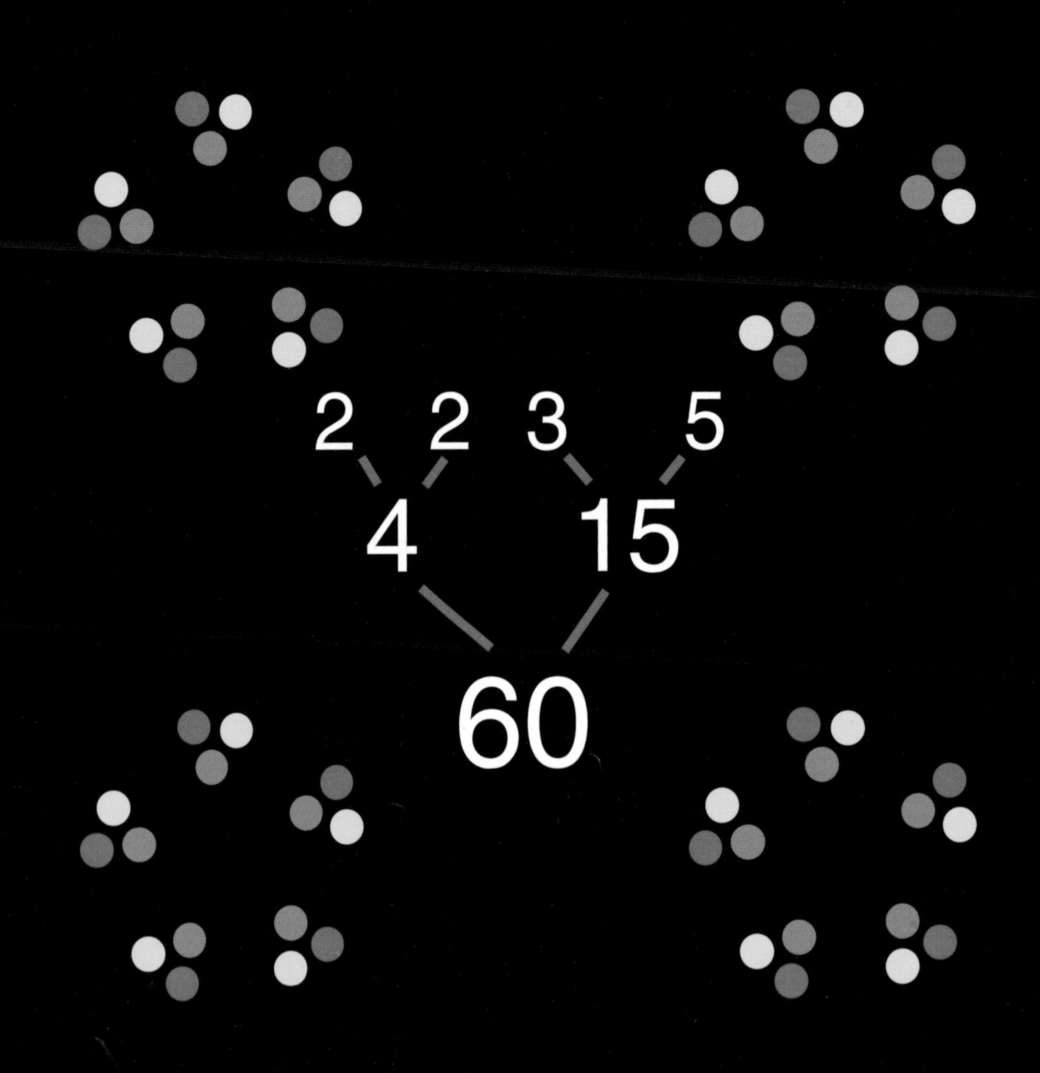
2
2
3
5
4
15
60

61

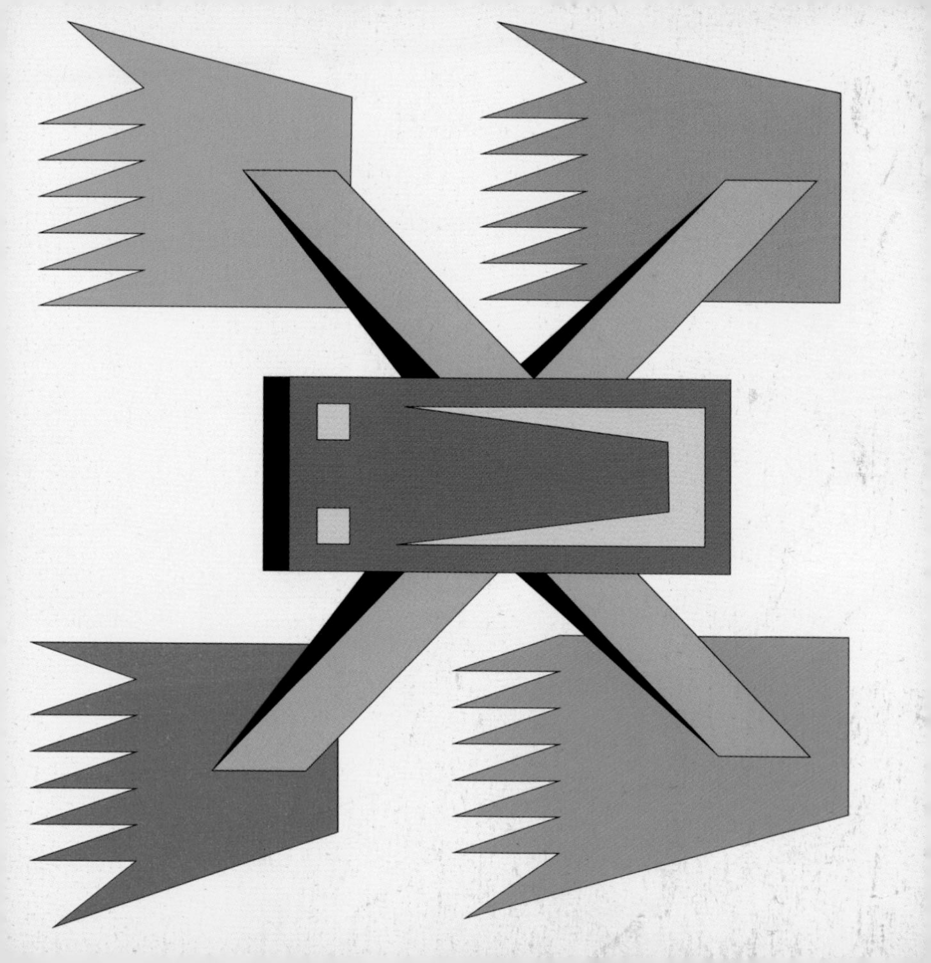

2
31
62

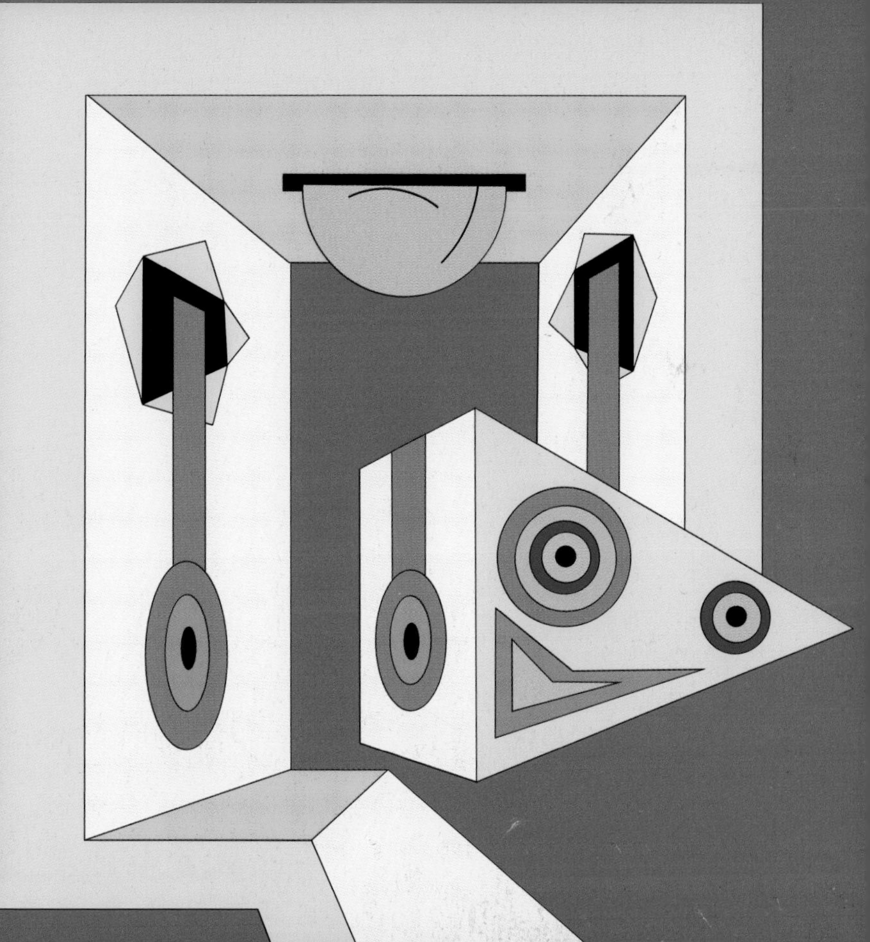

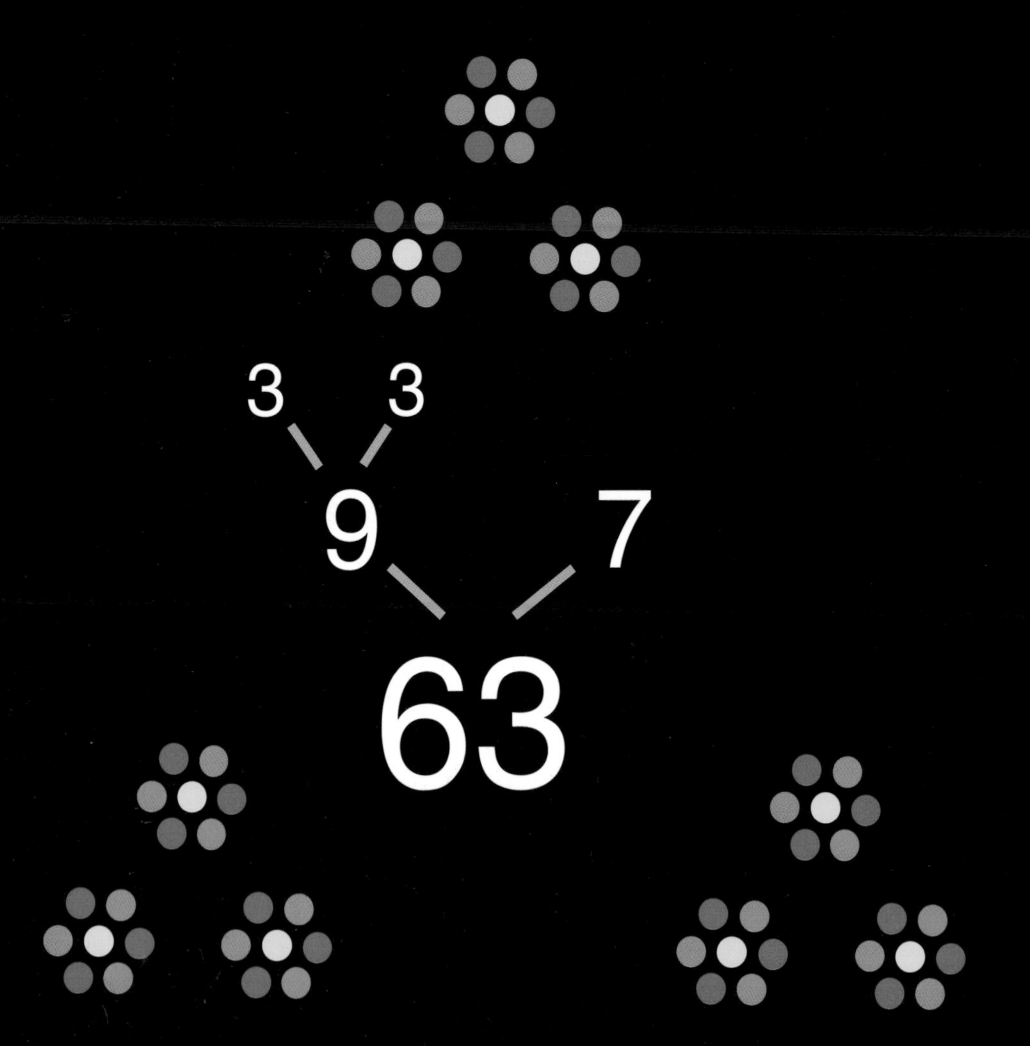
3
3
9
7
63

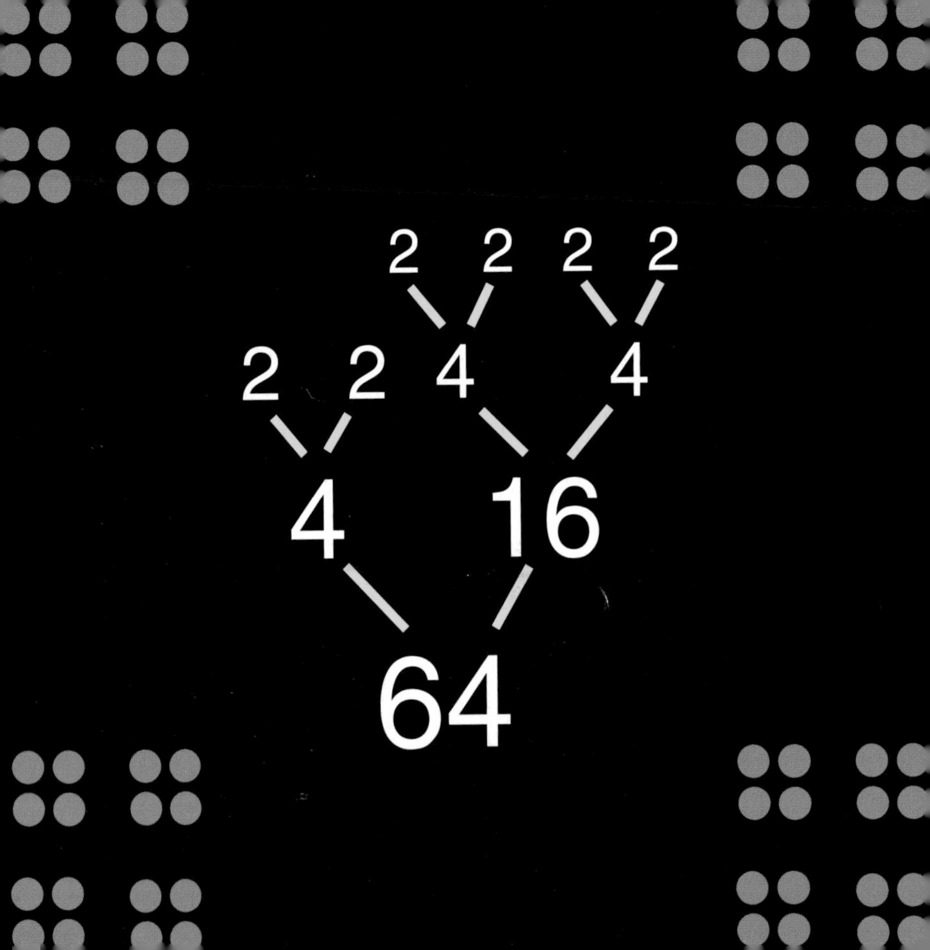
2 2 2 2
2 2 4 4
4 16
64

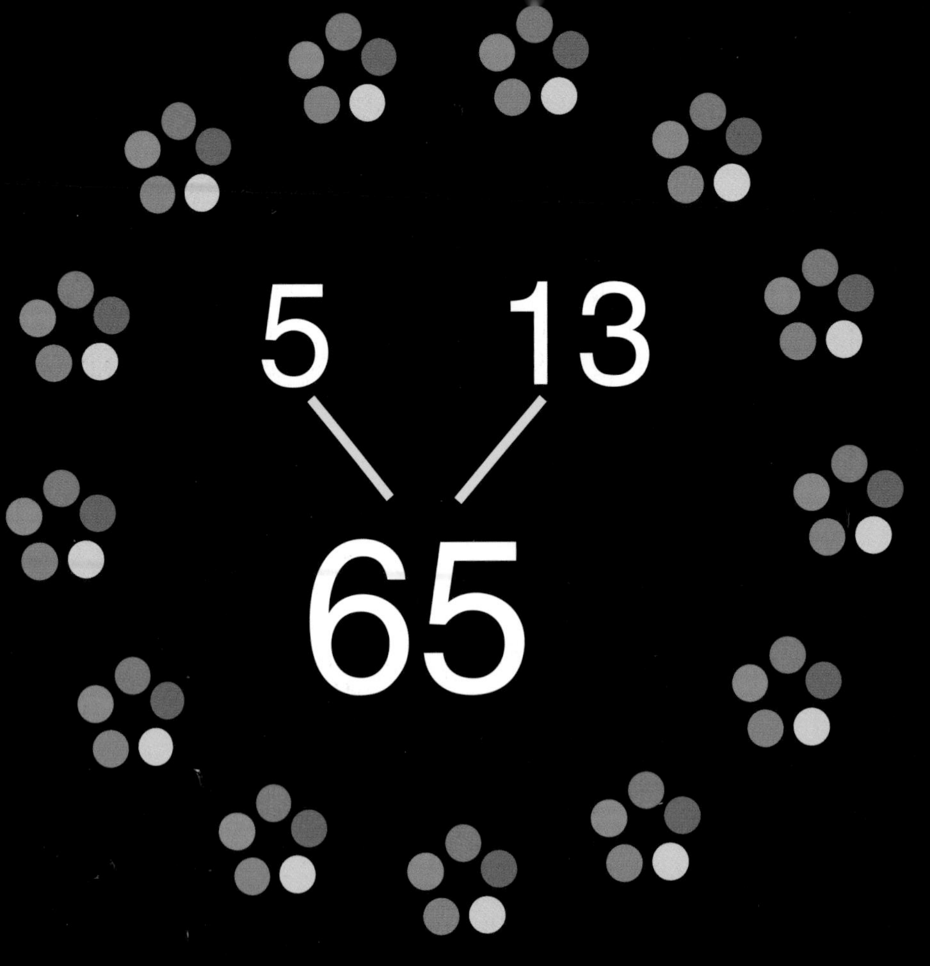
5
13
65

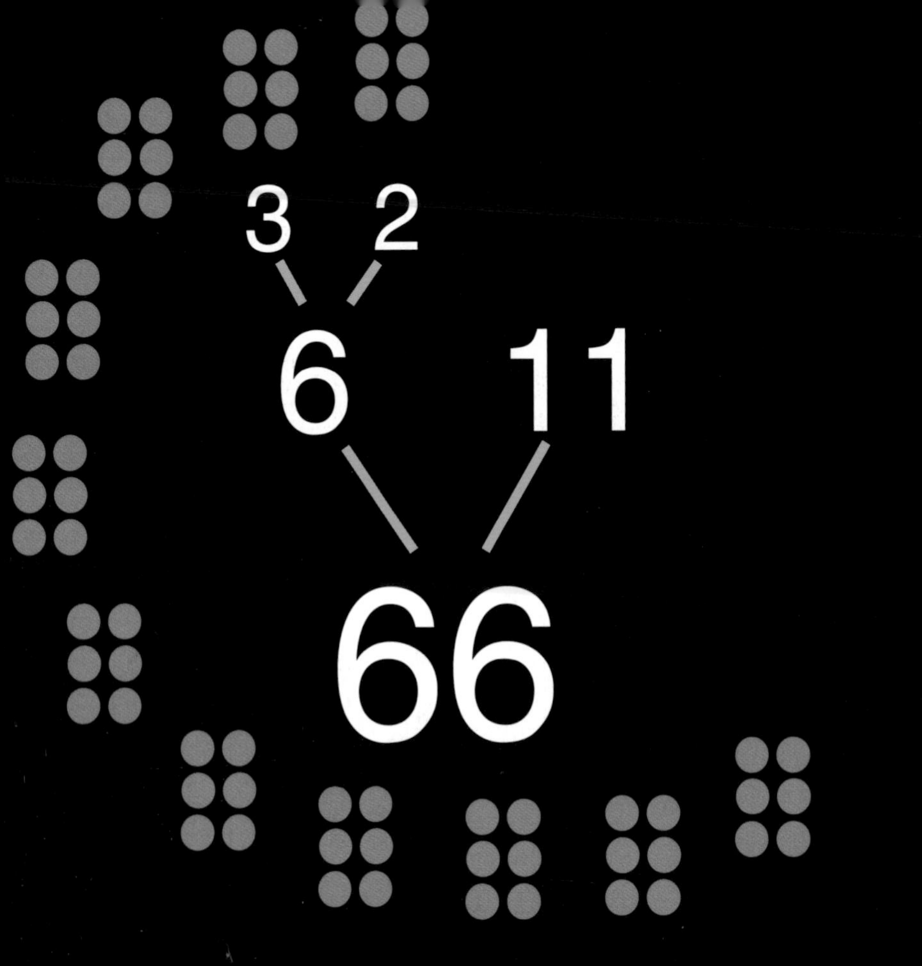

3
2
6
11
66

67

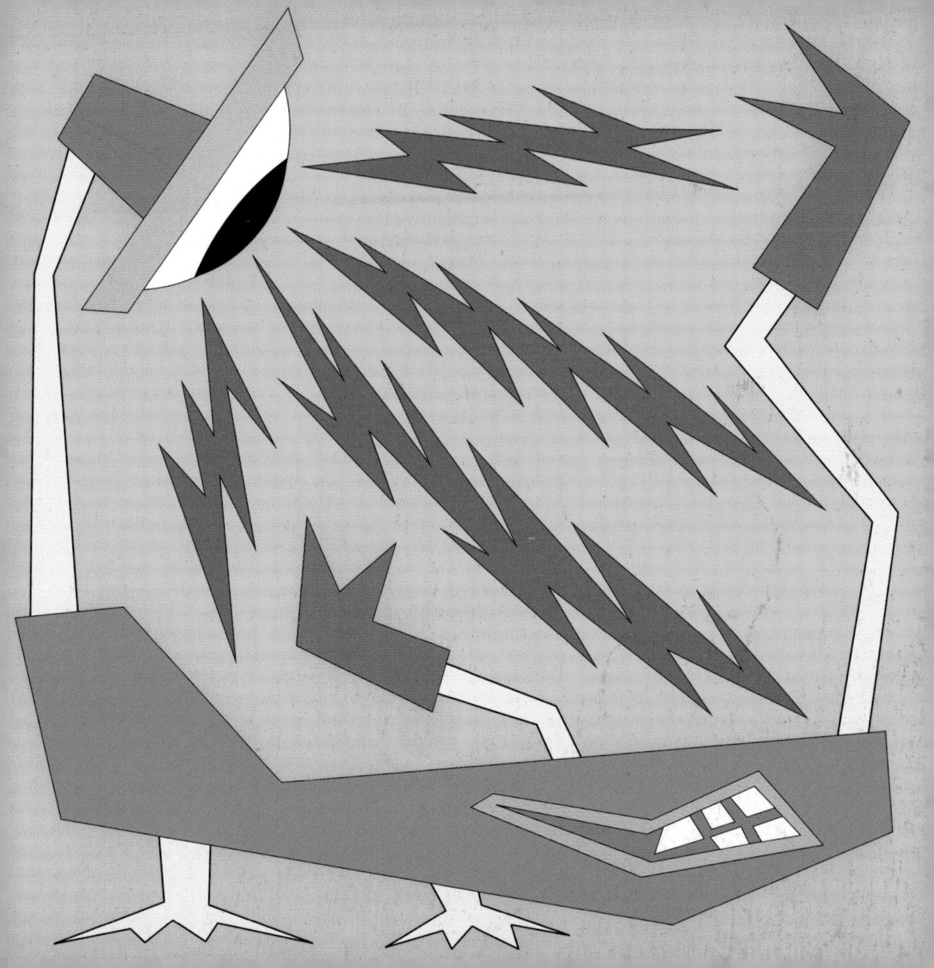

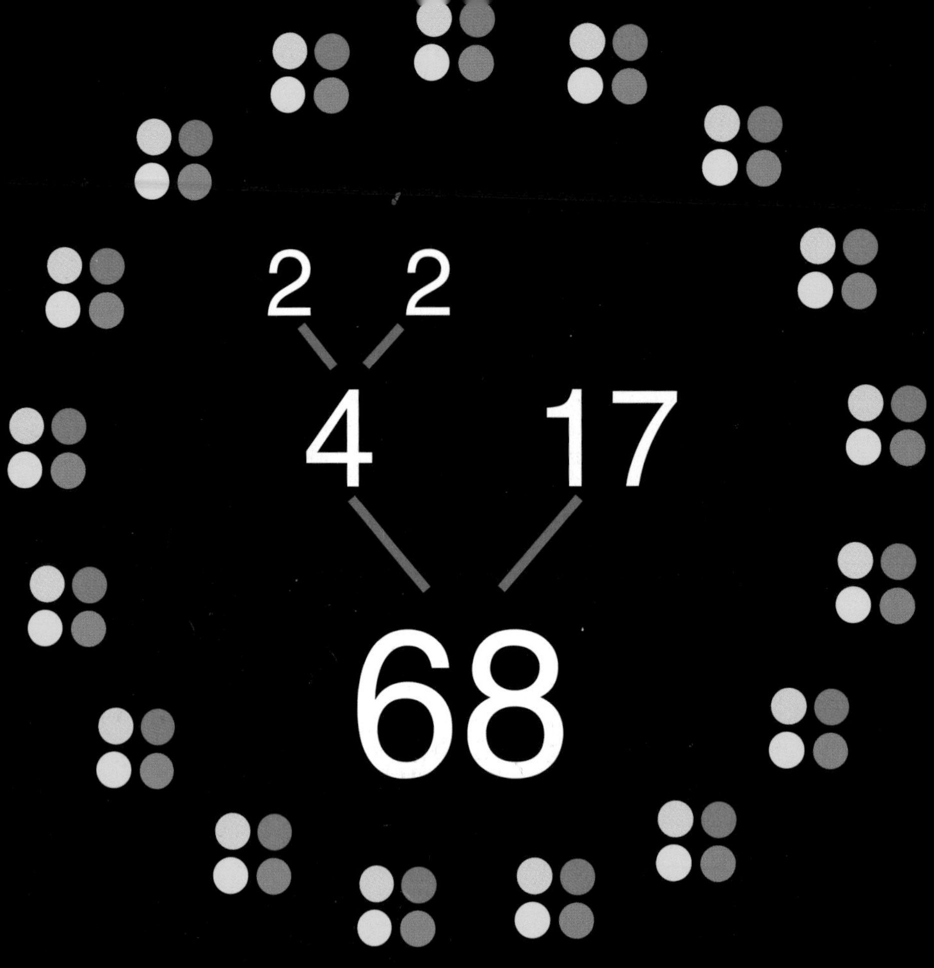
2
2
4
17
68

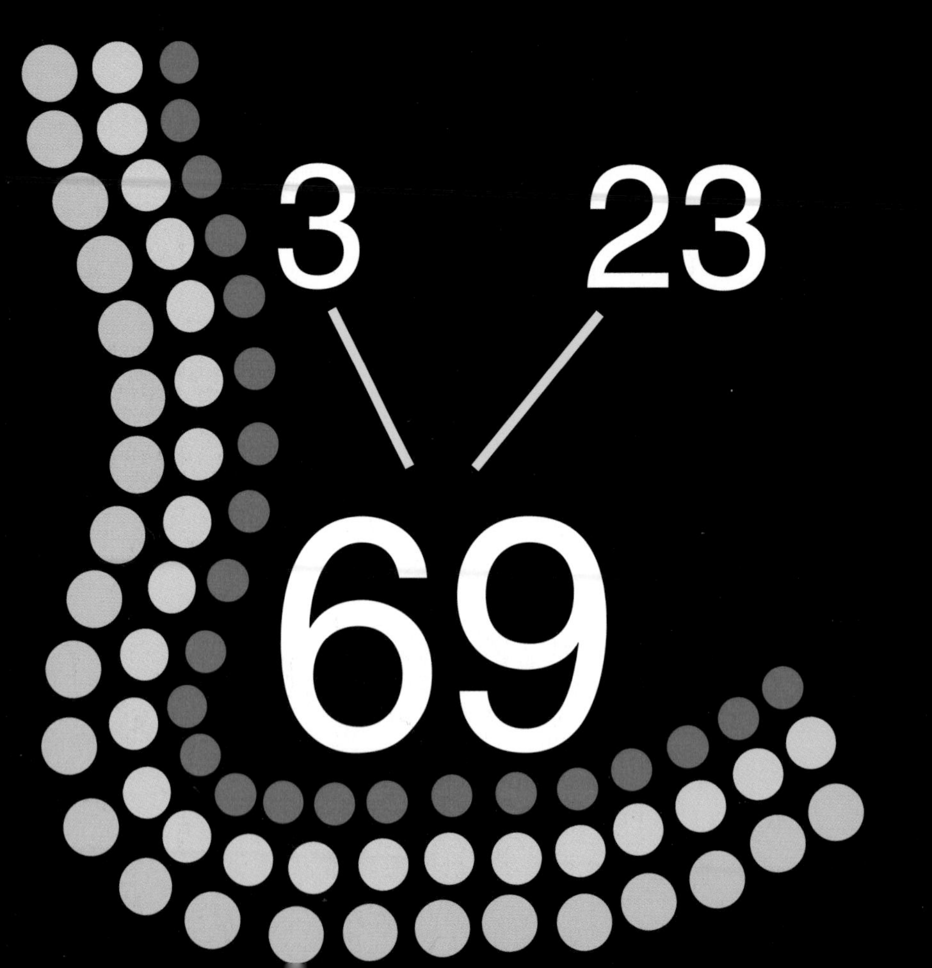
3
23
69

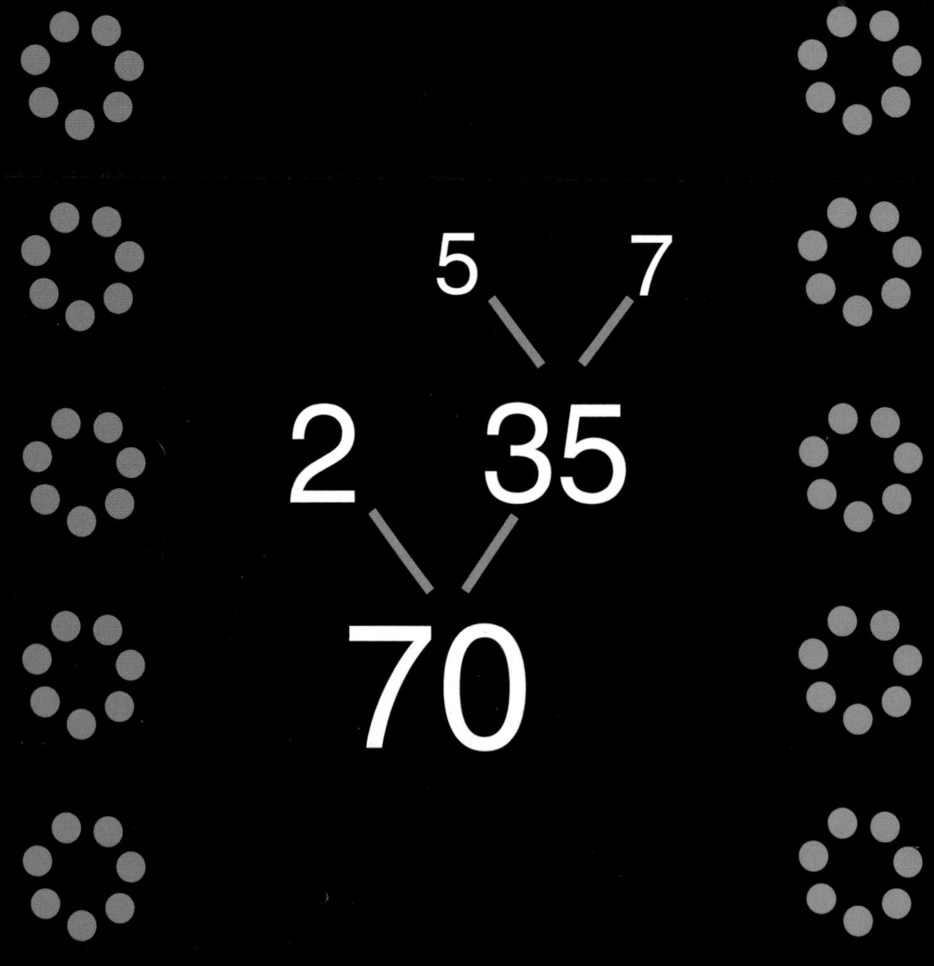
5
7
2
35
70

71

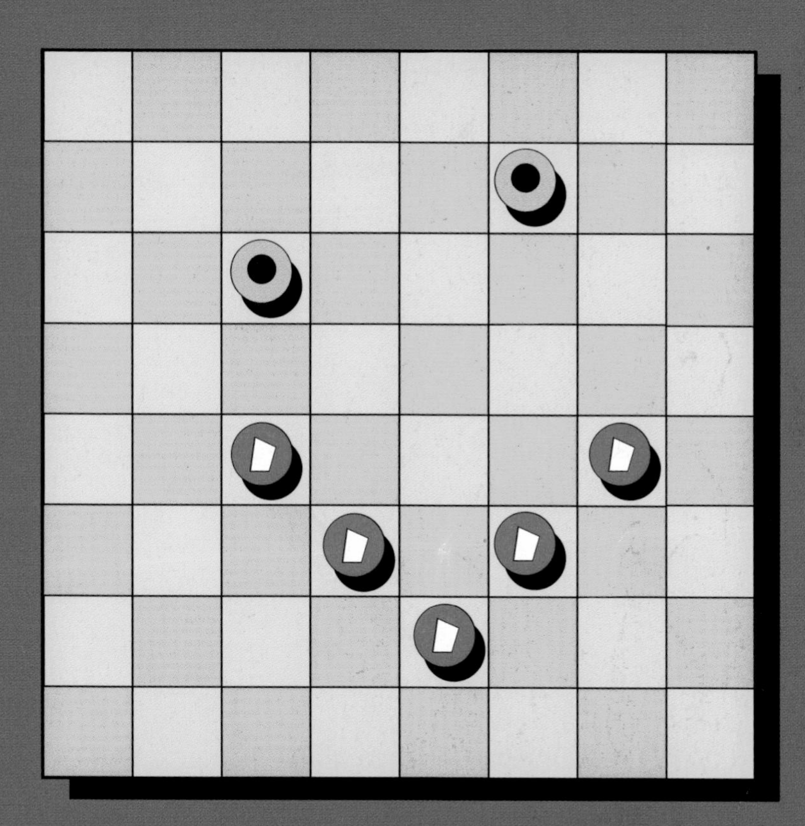

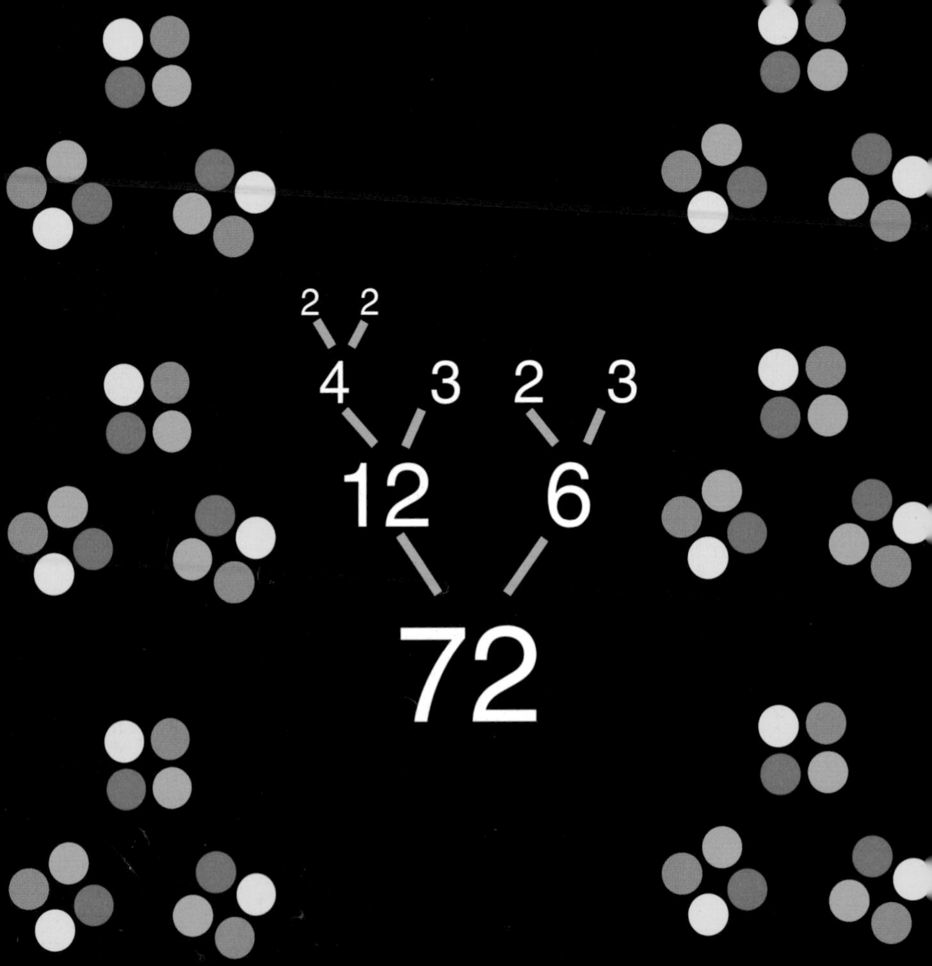
2 2
4 3 2 3
12 6
72

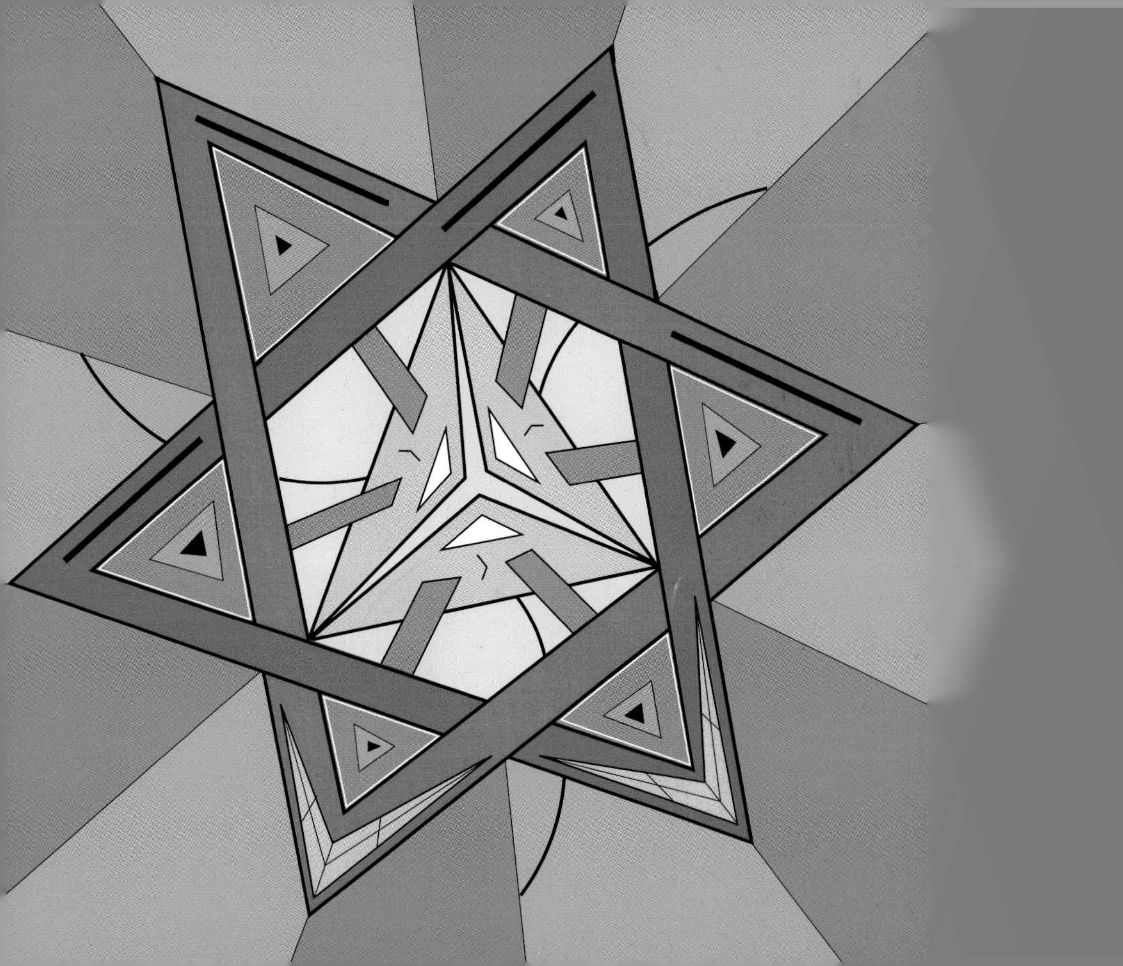

73

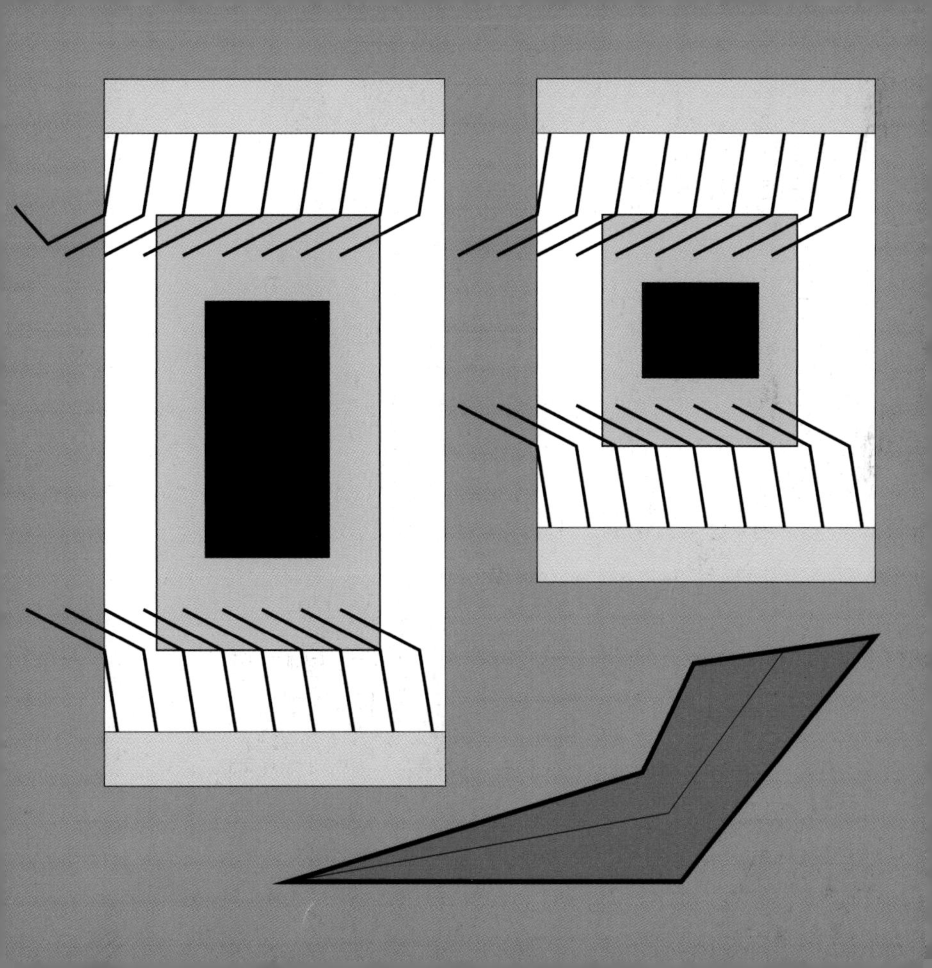

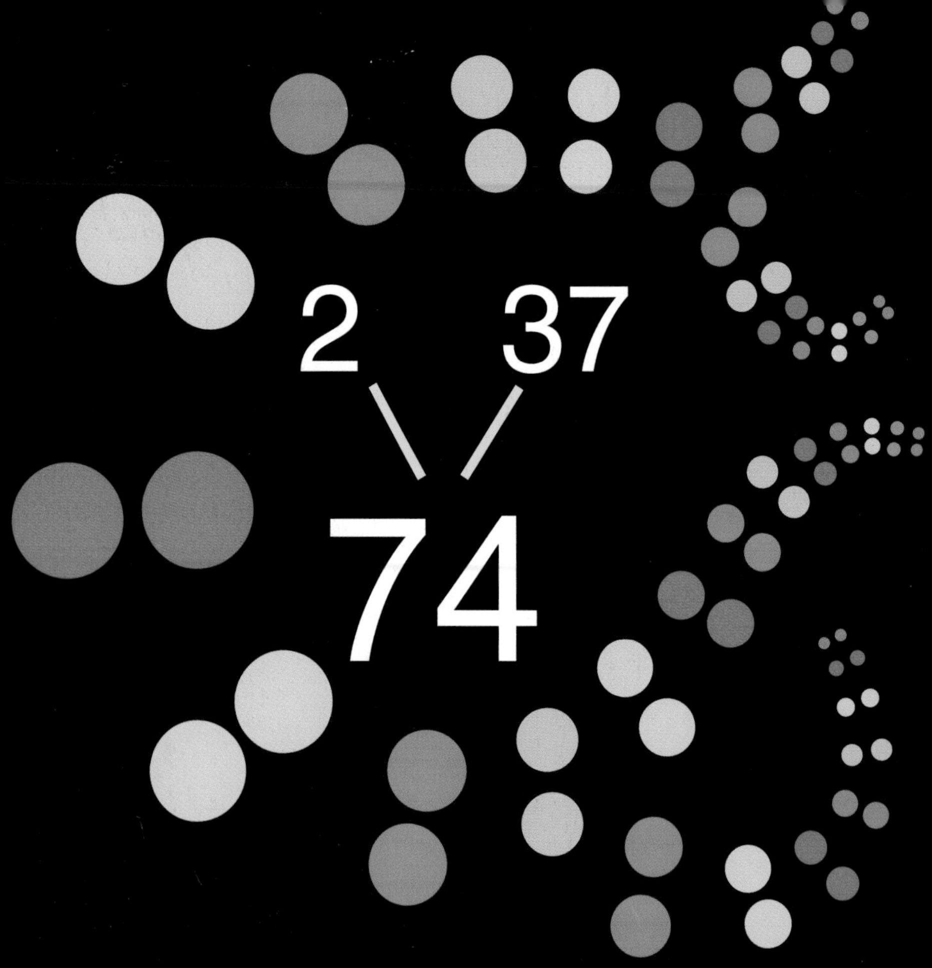
2
37
74

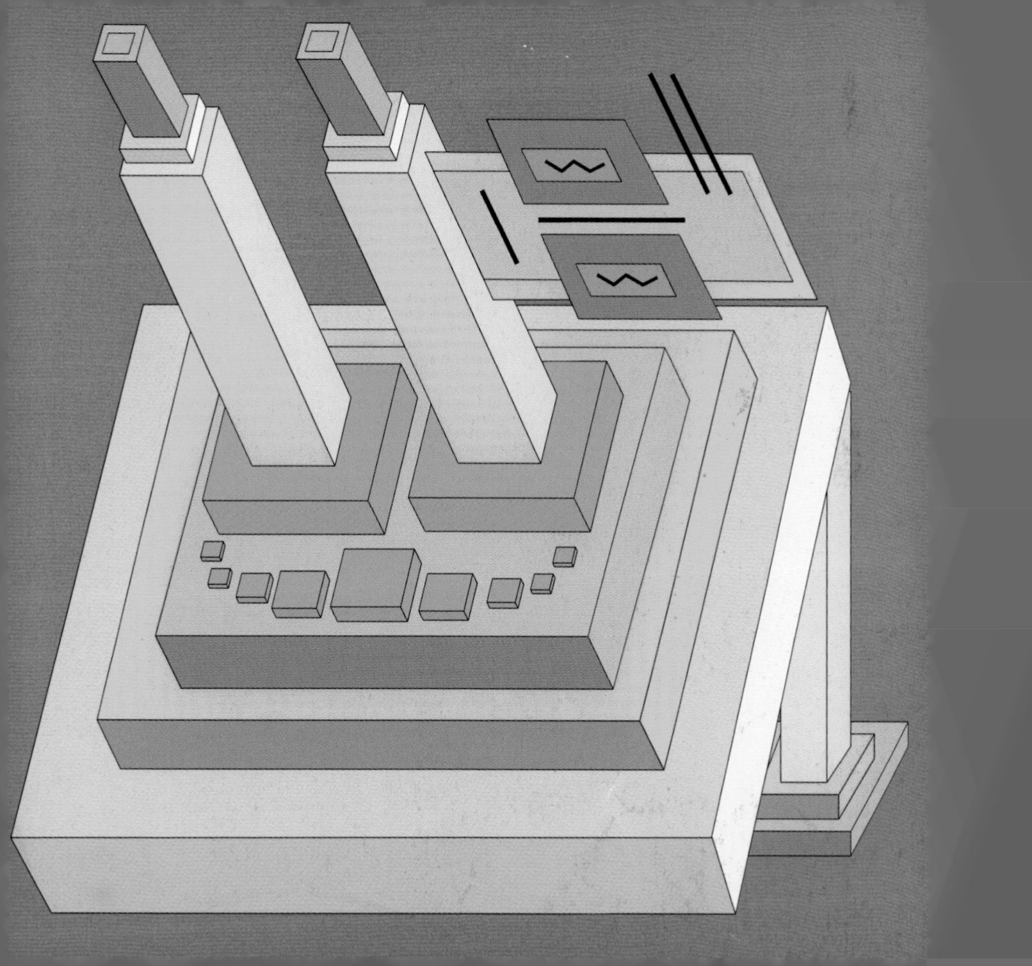

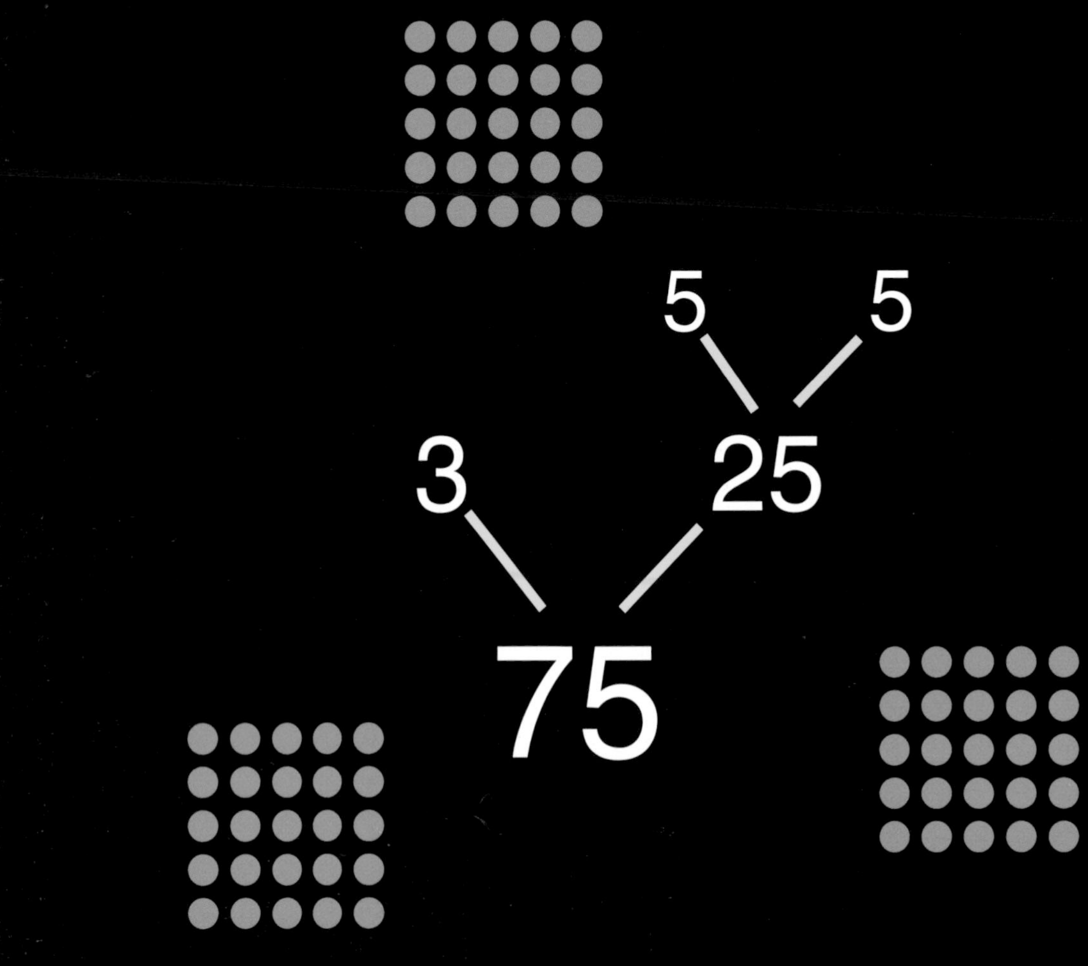
5
5
3
25
75

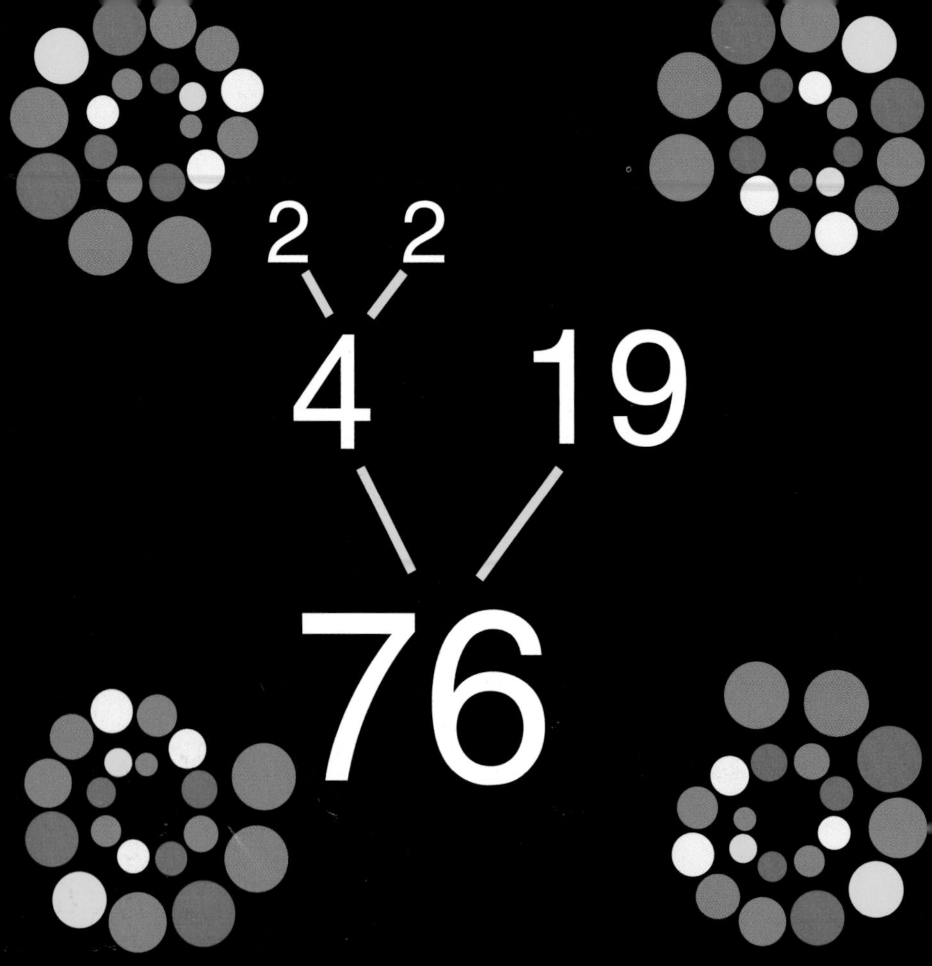
2
2
4
19
76

7
11
77

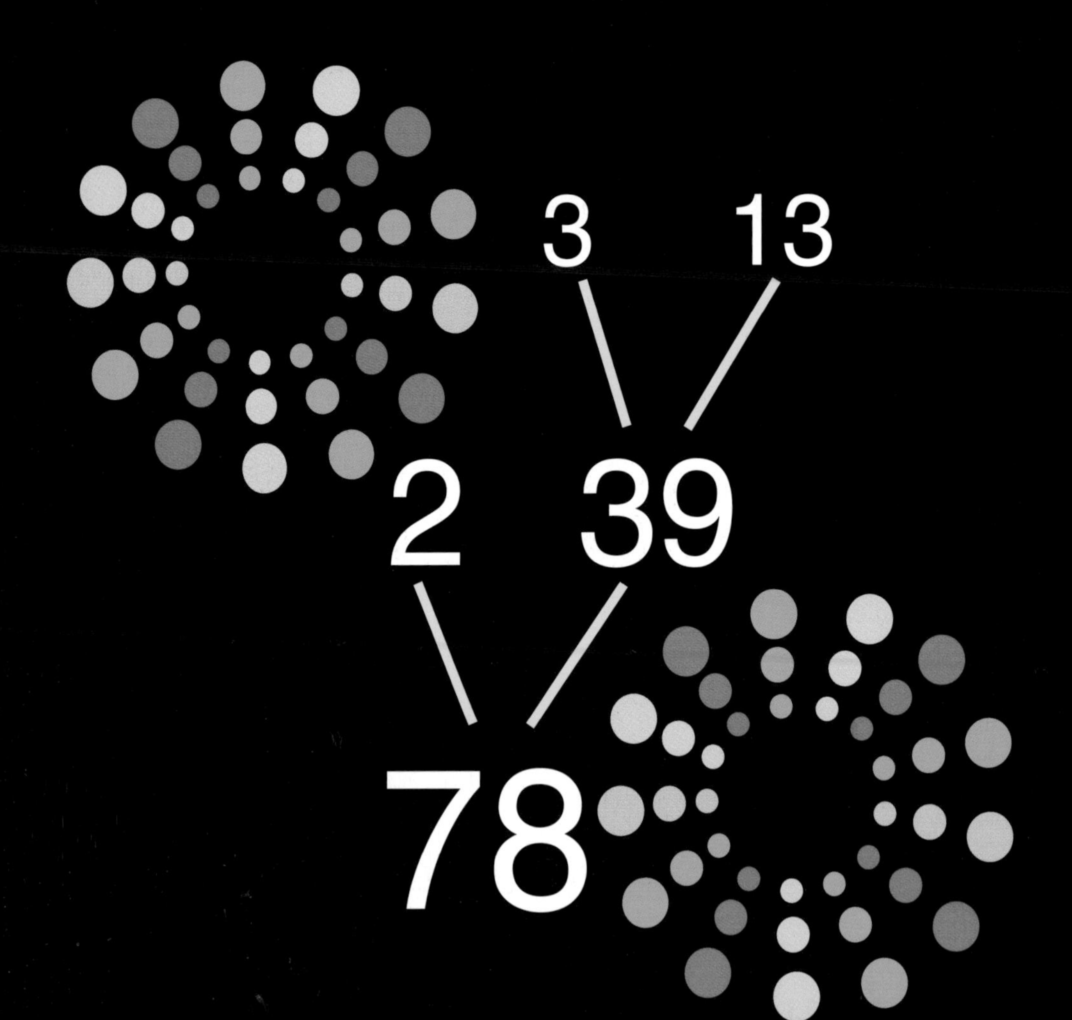
3
13
2
39
78

79

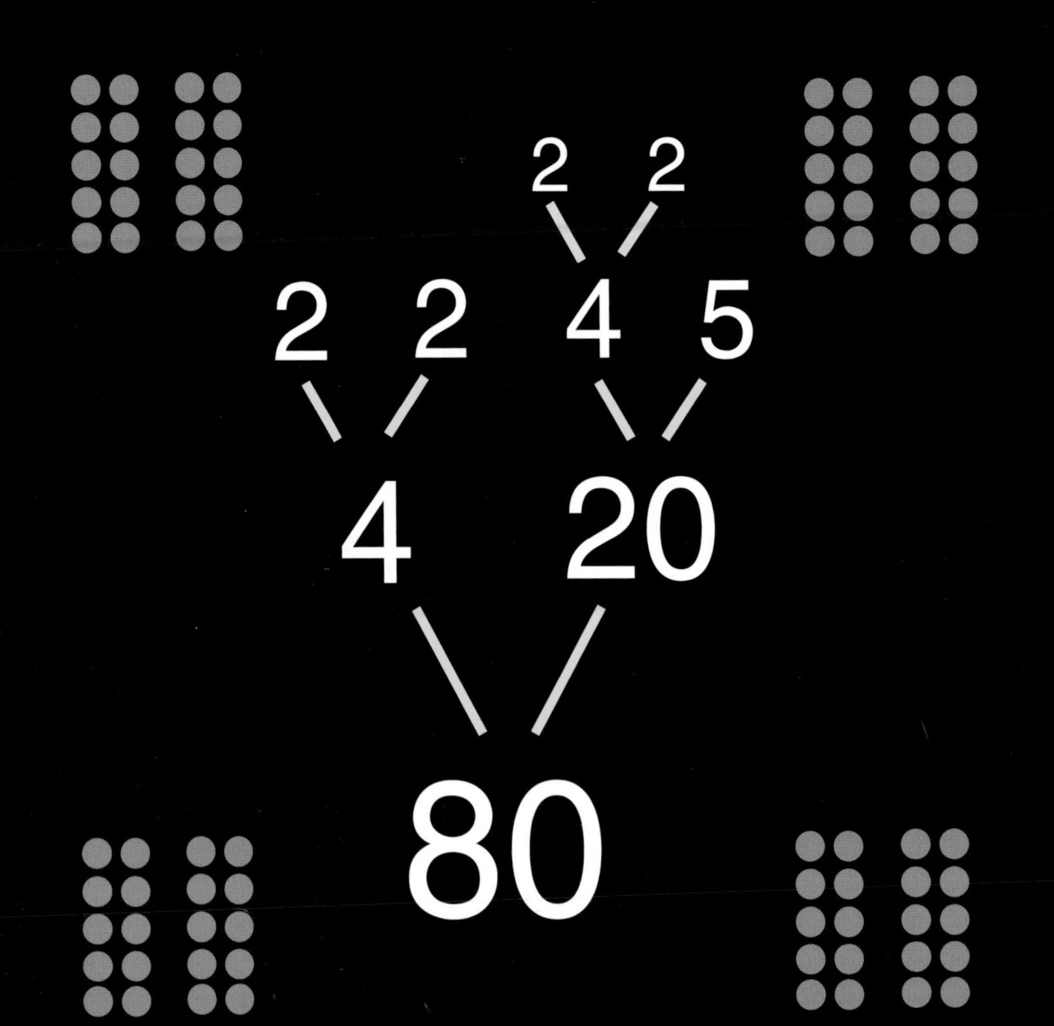
2 2
2 2 4 5
4 20
80

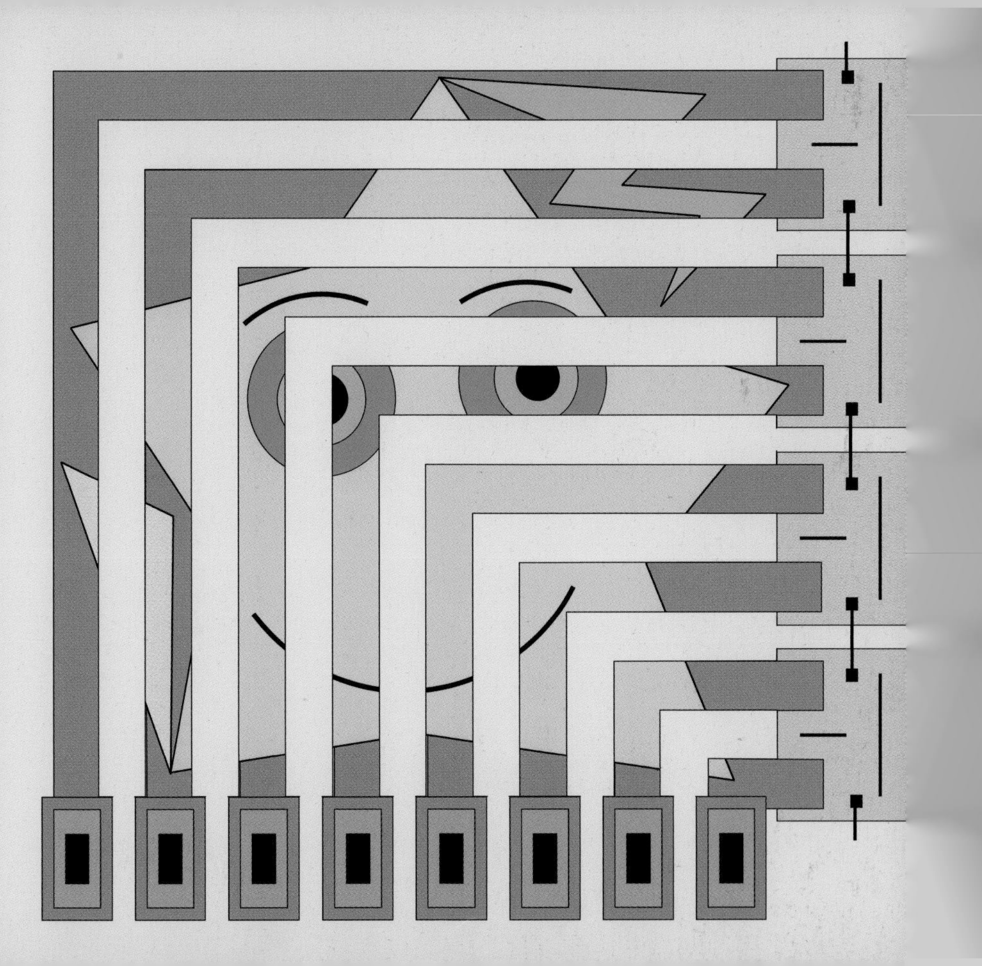

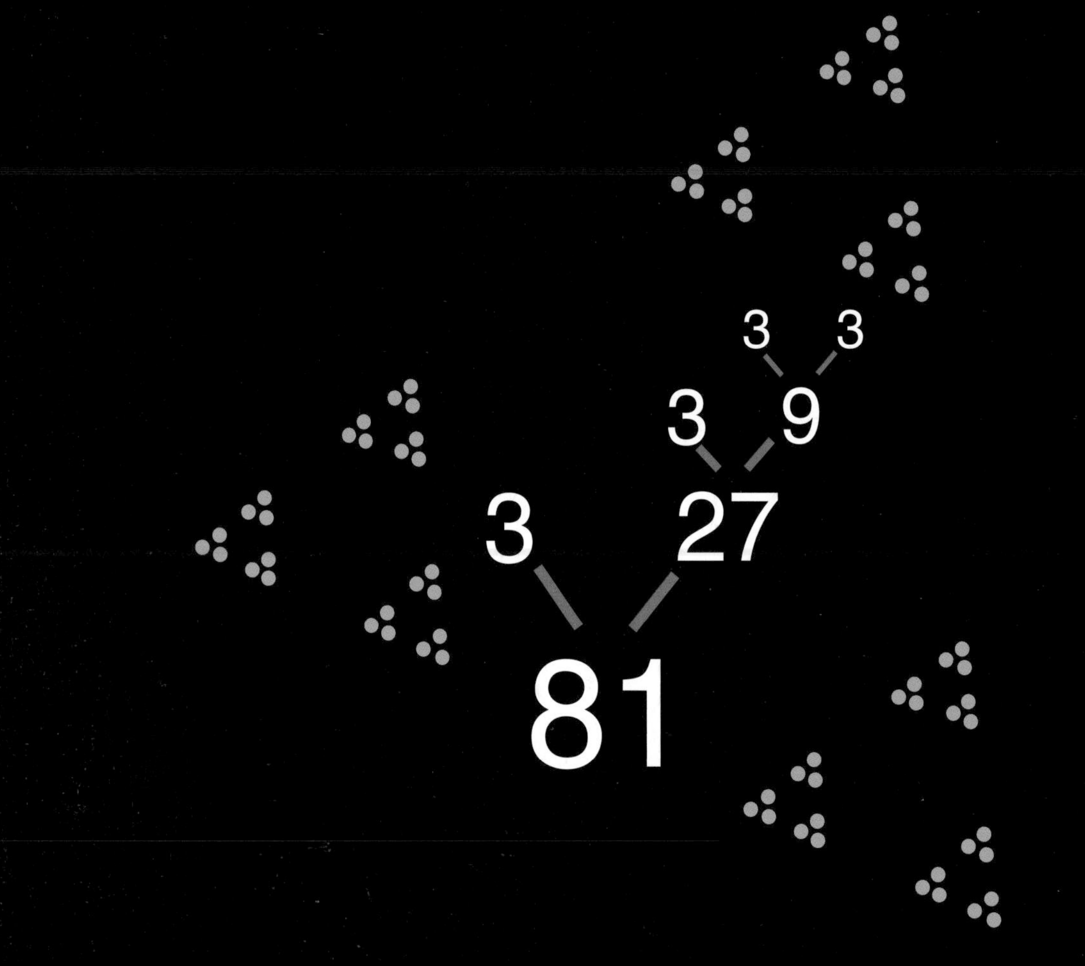
3
3
3
9
3
27
81

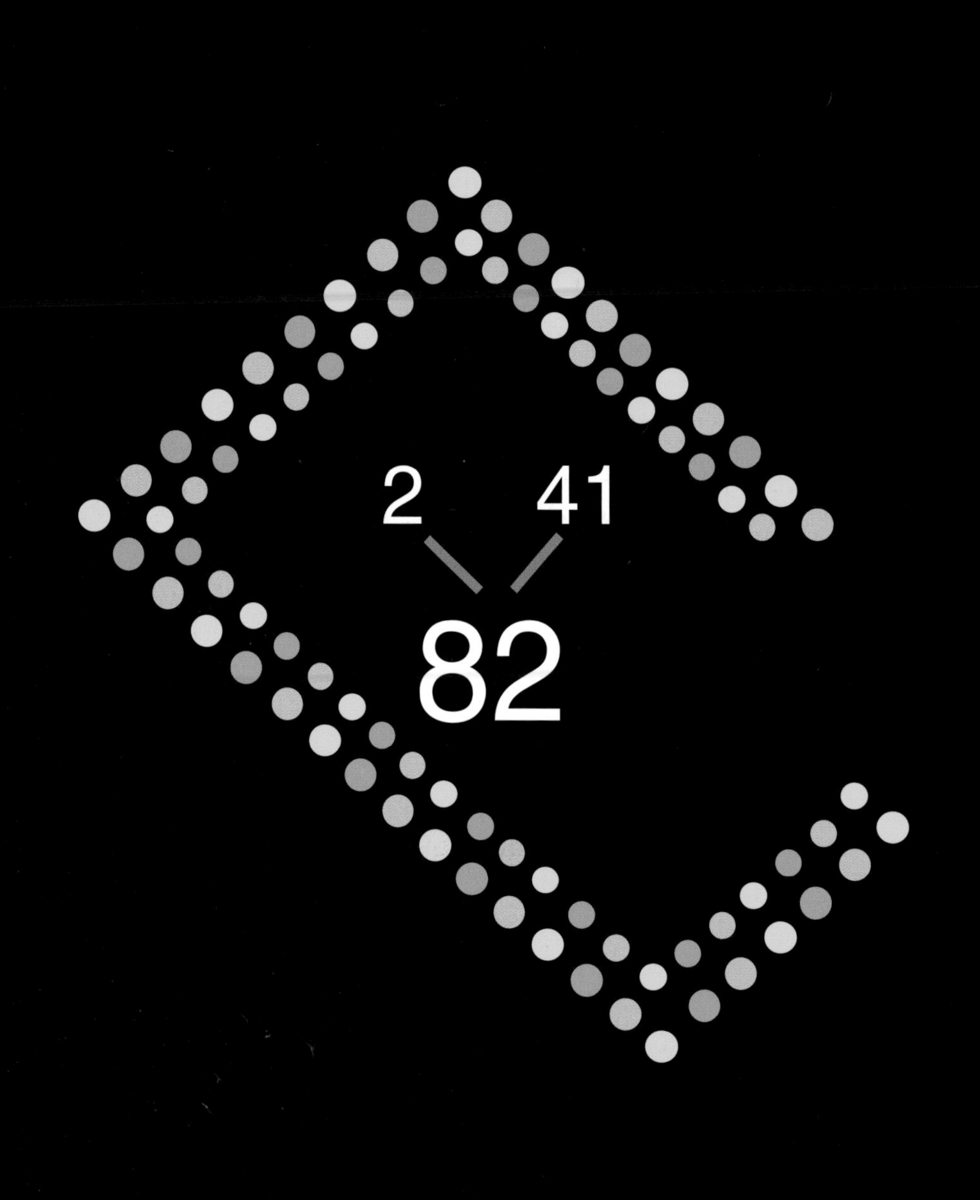
2
41
82

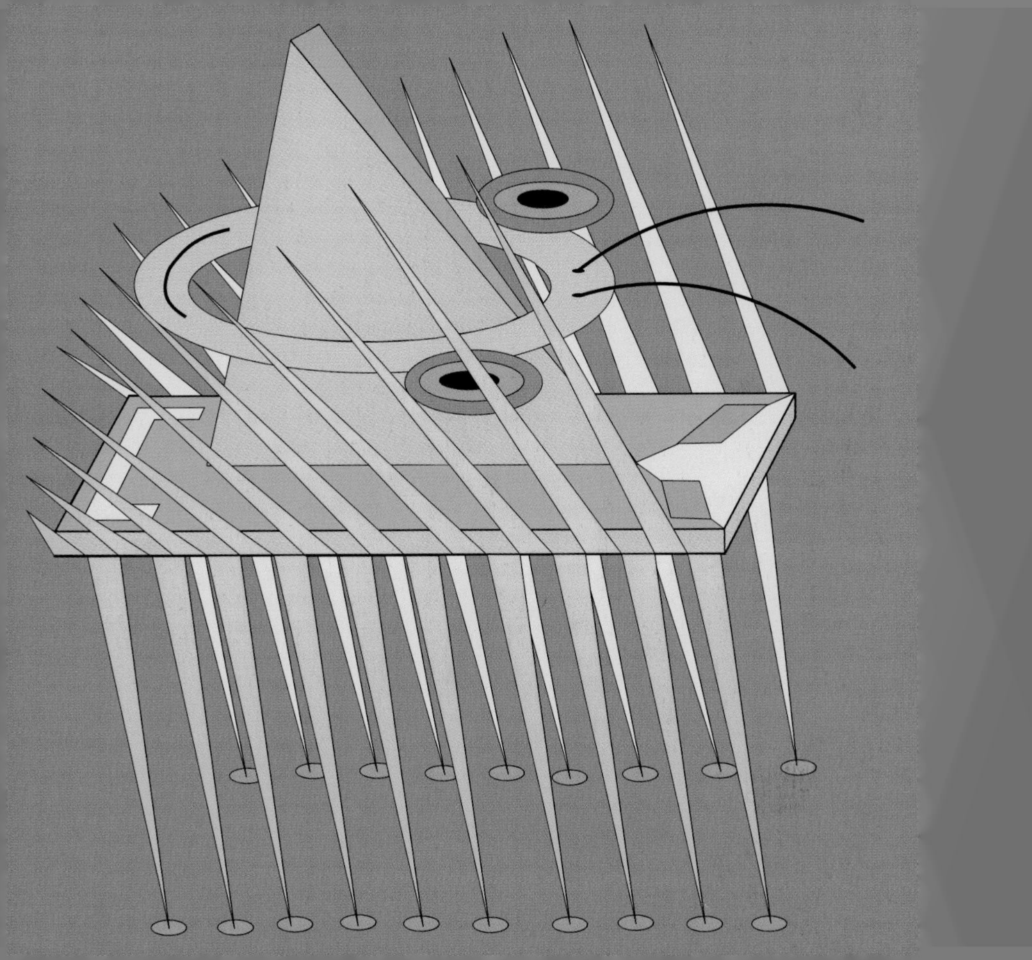

83

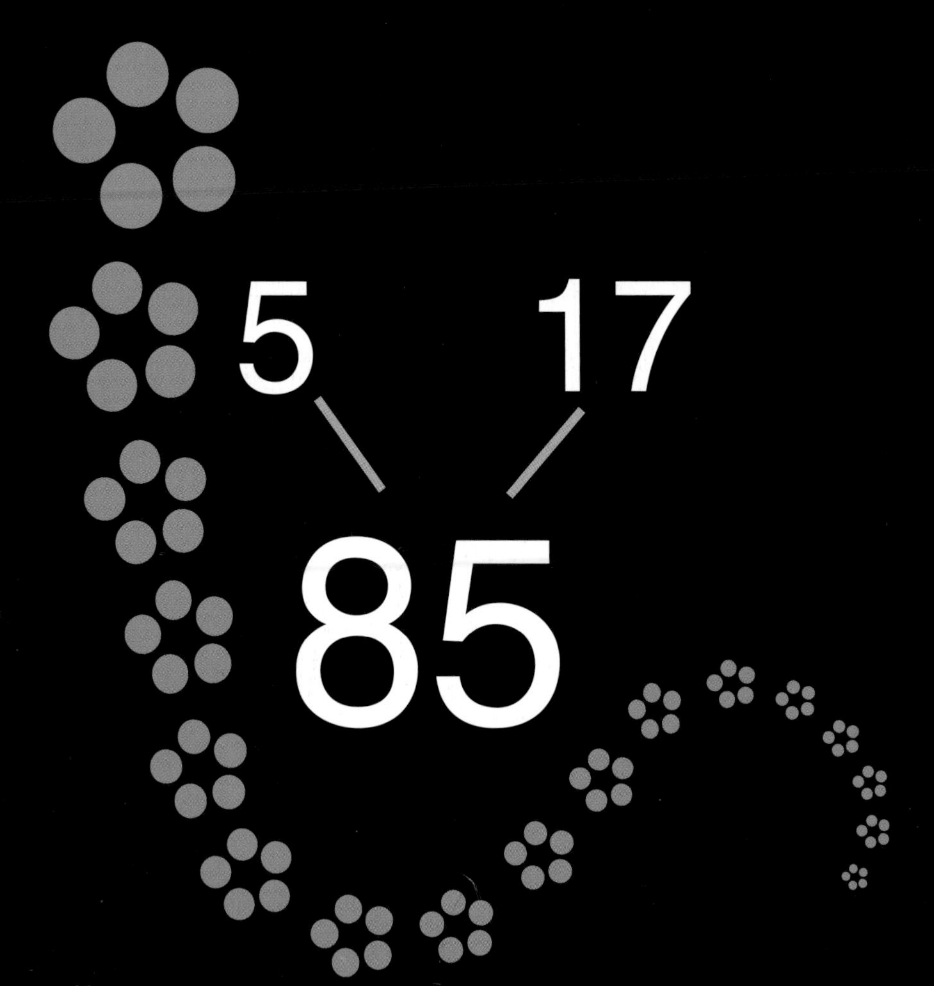
5
17
85

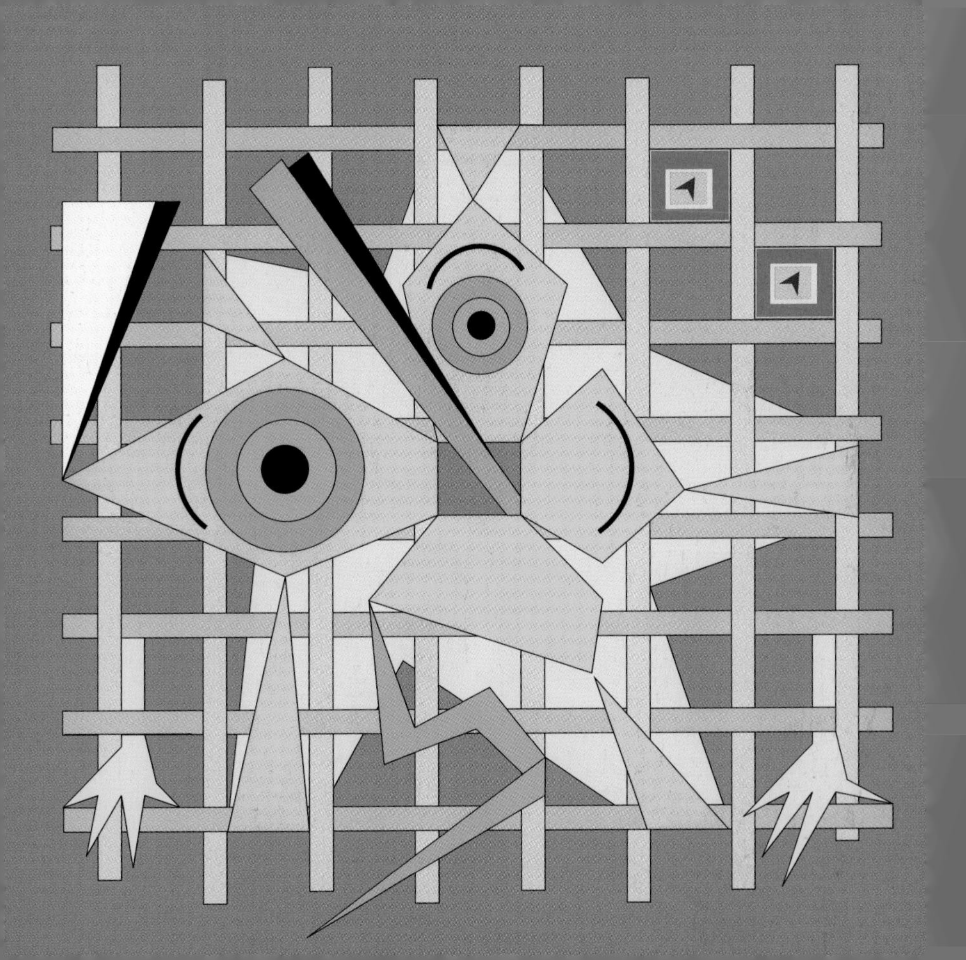

2 43

86

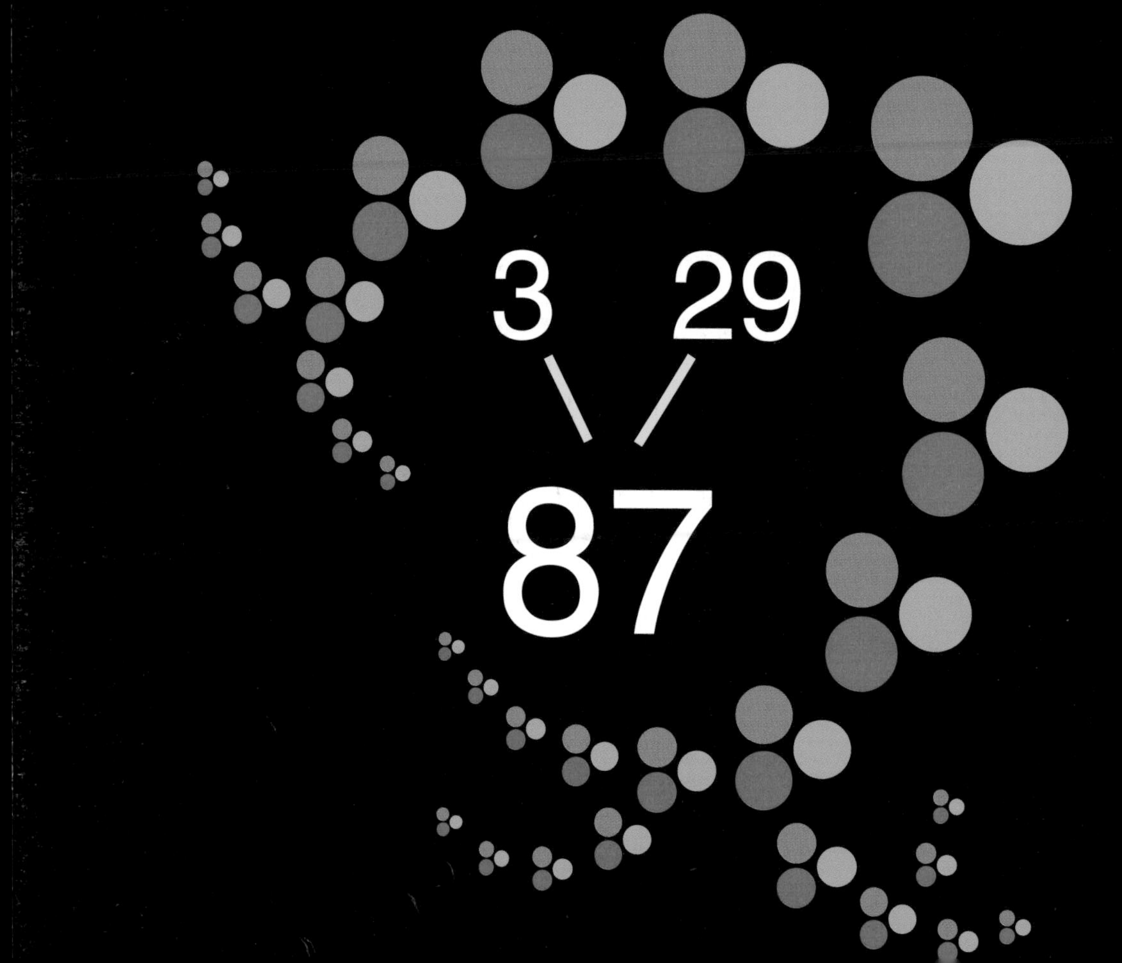
3
29
87

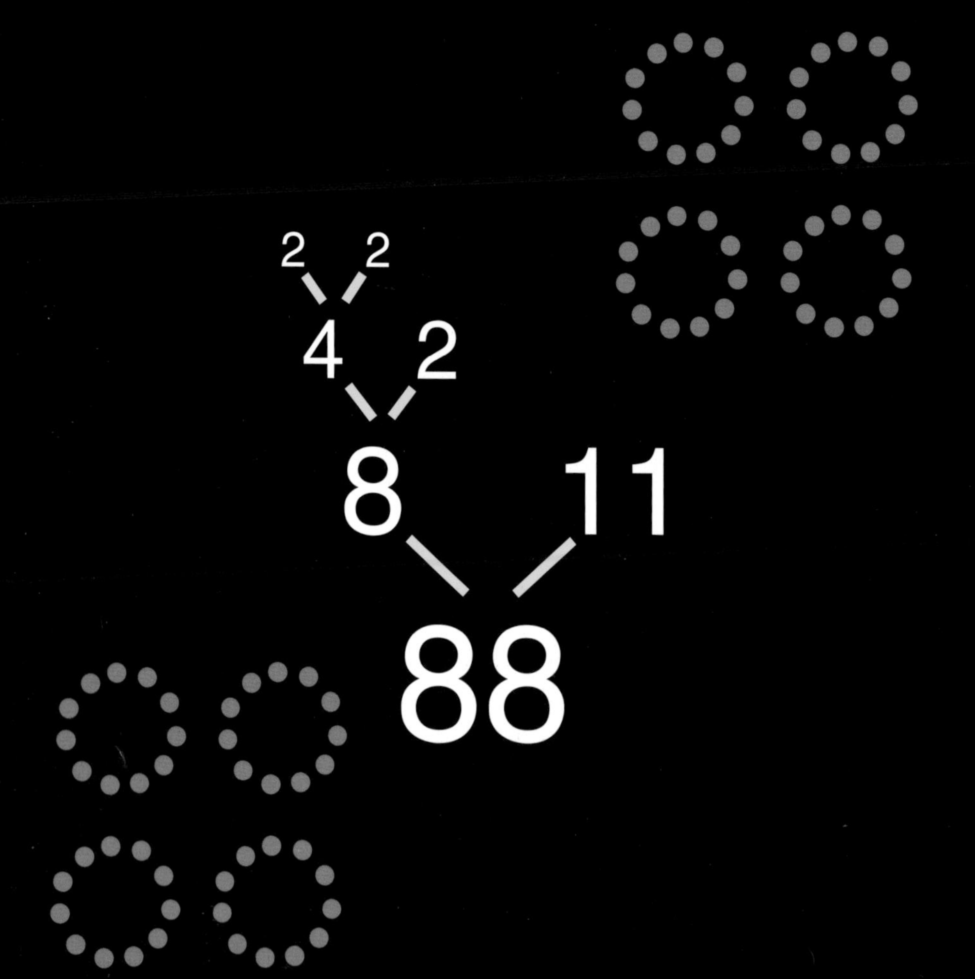
2
2
4
2
8
11
88

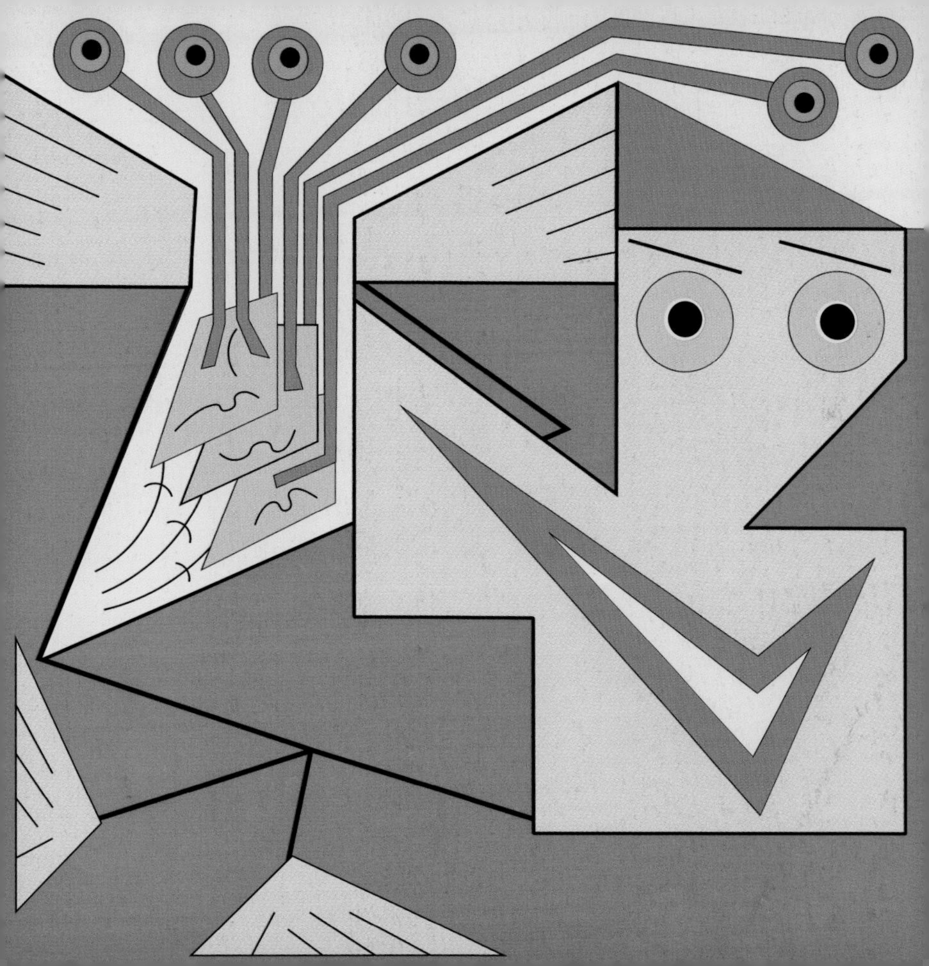

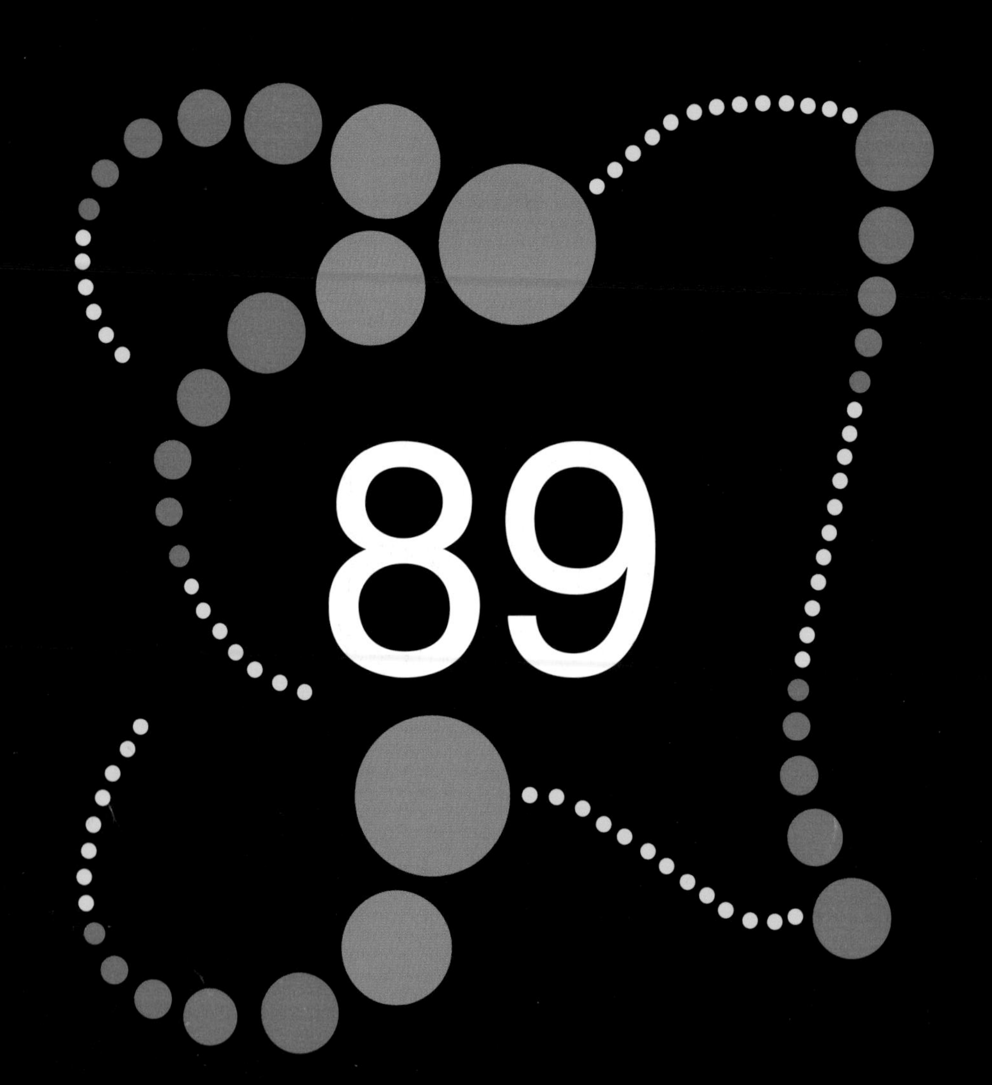
89

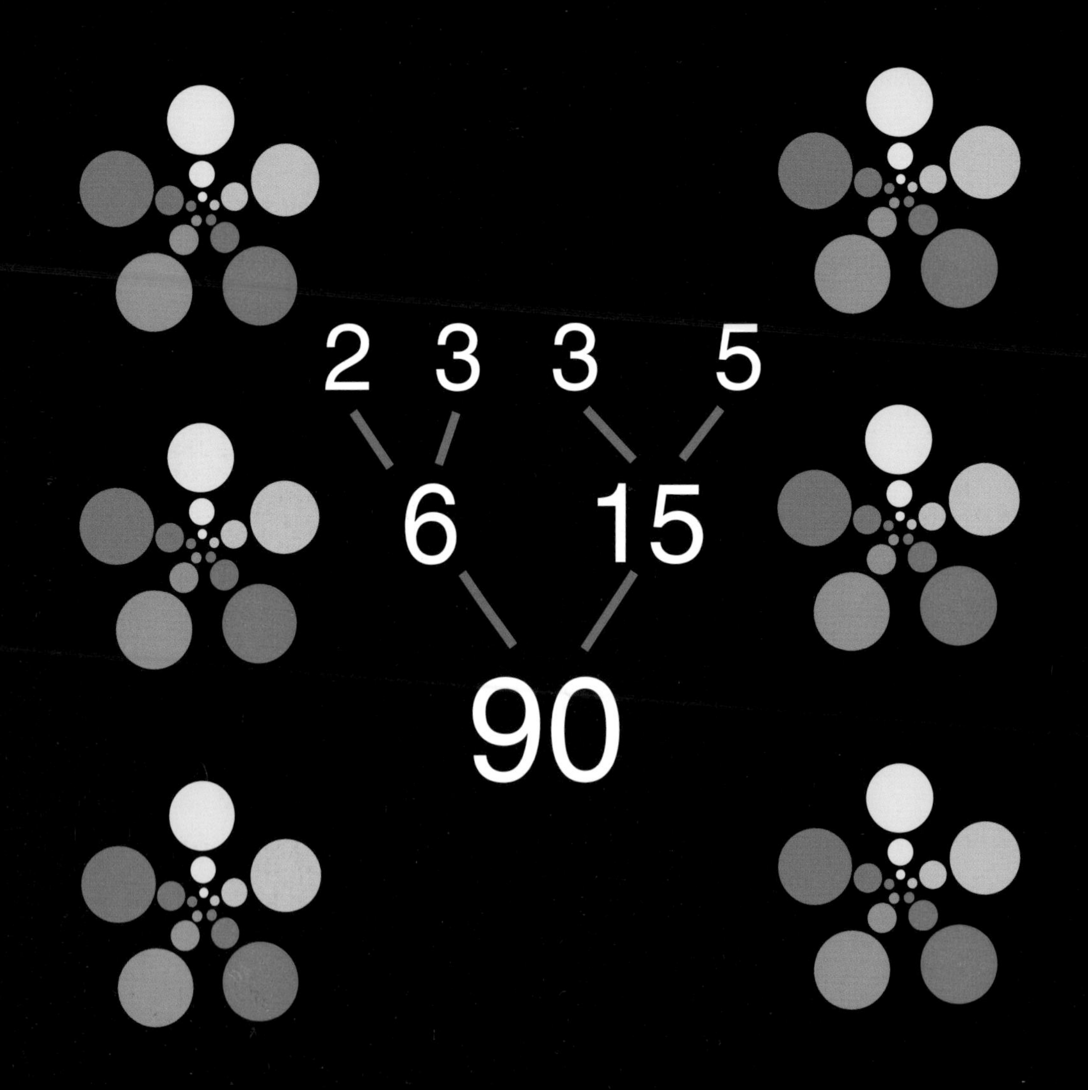
2 3 3 5
6 15
90

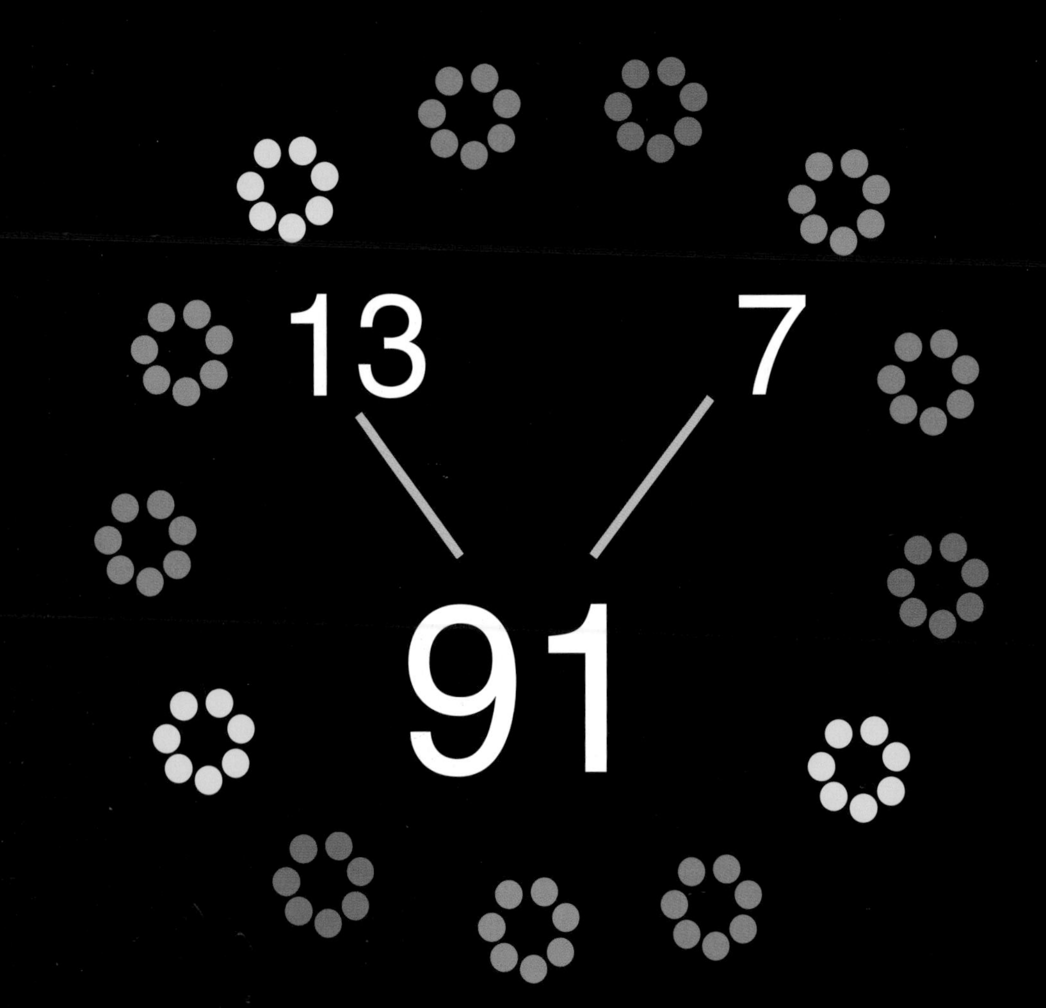
13
7
91

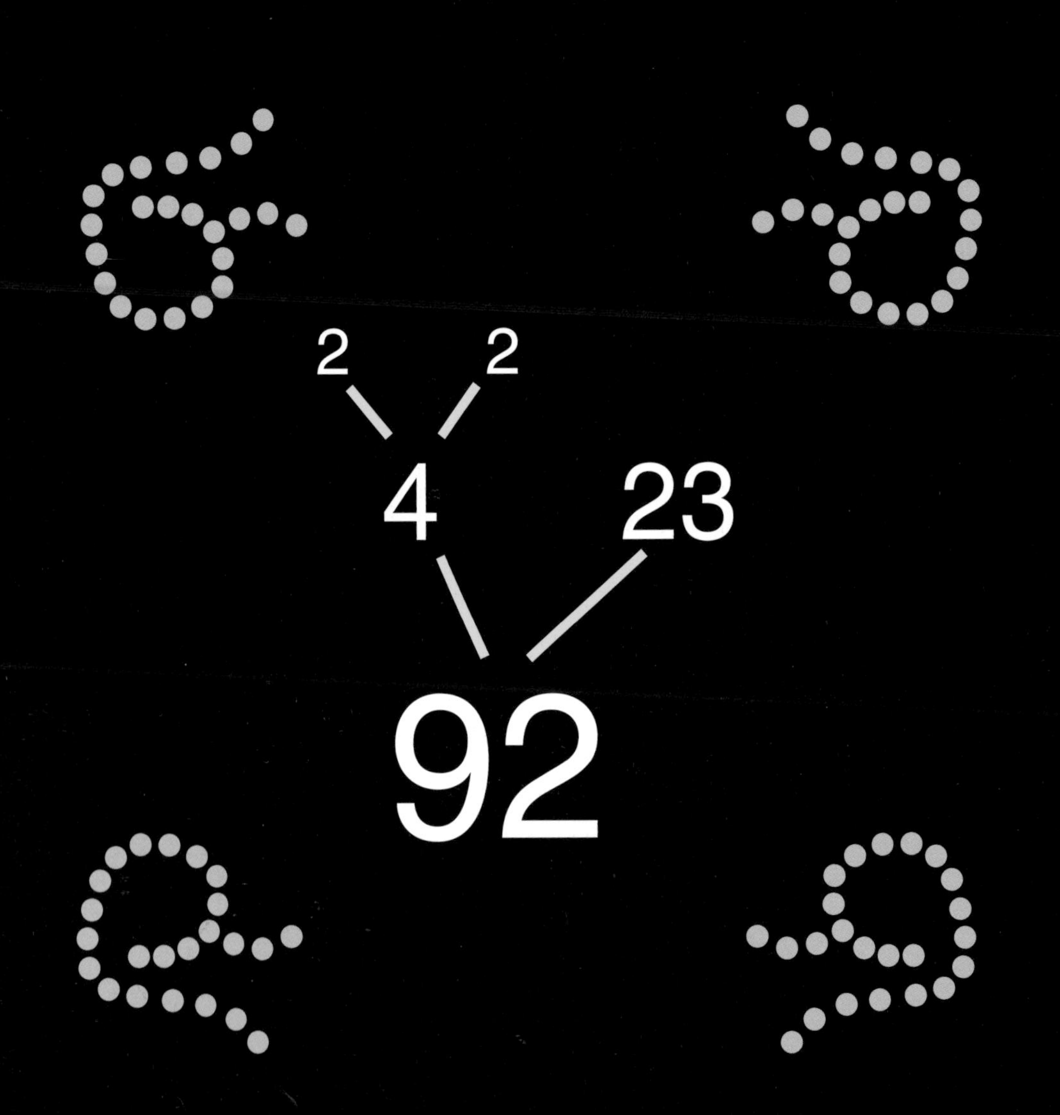
2
2
4
23
92

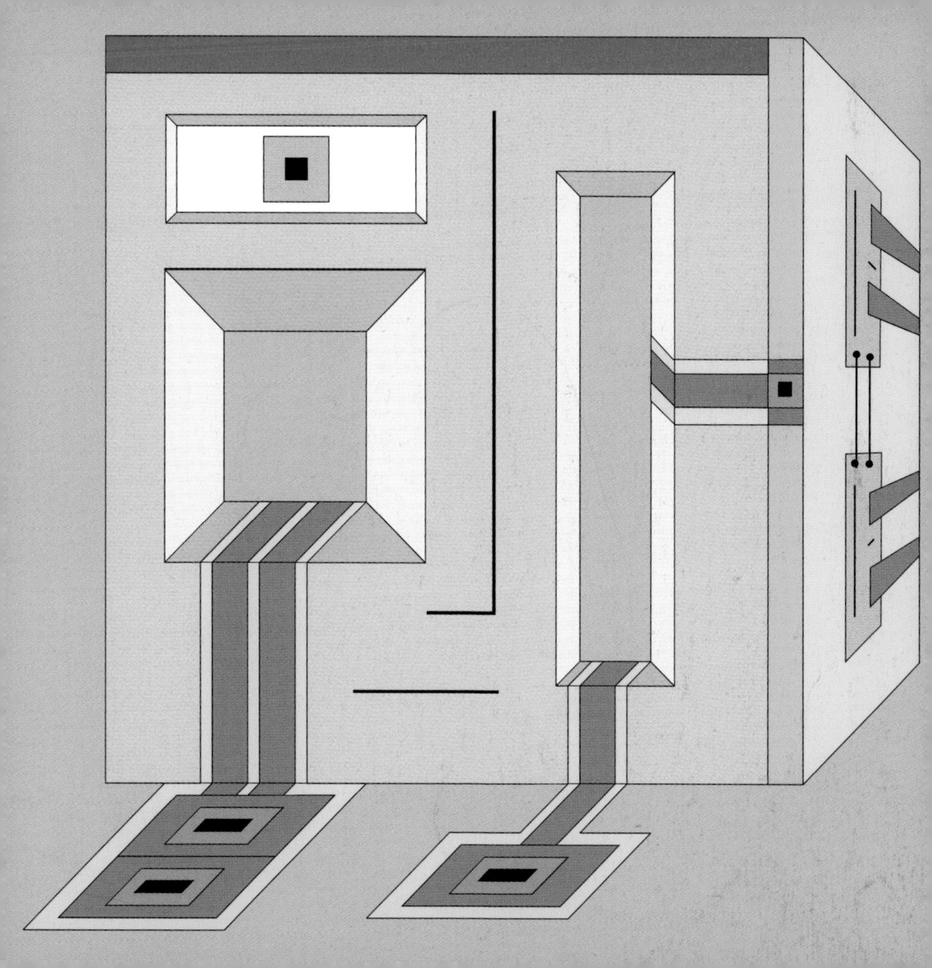

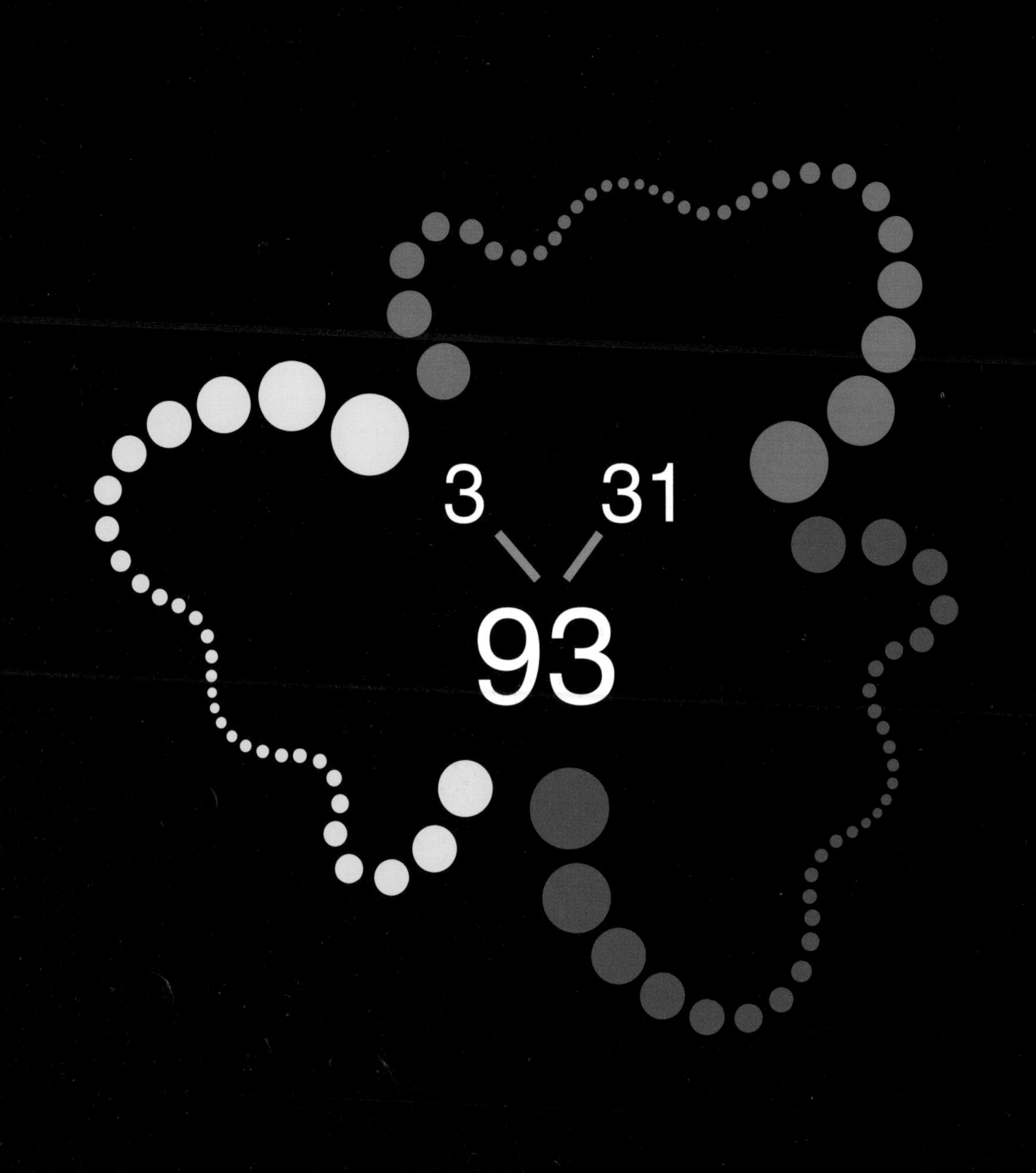
3
31
93

2
47
94

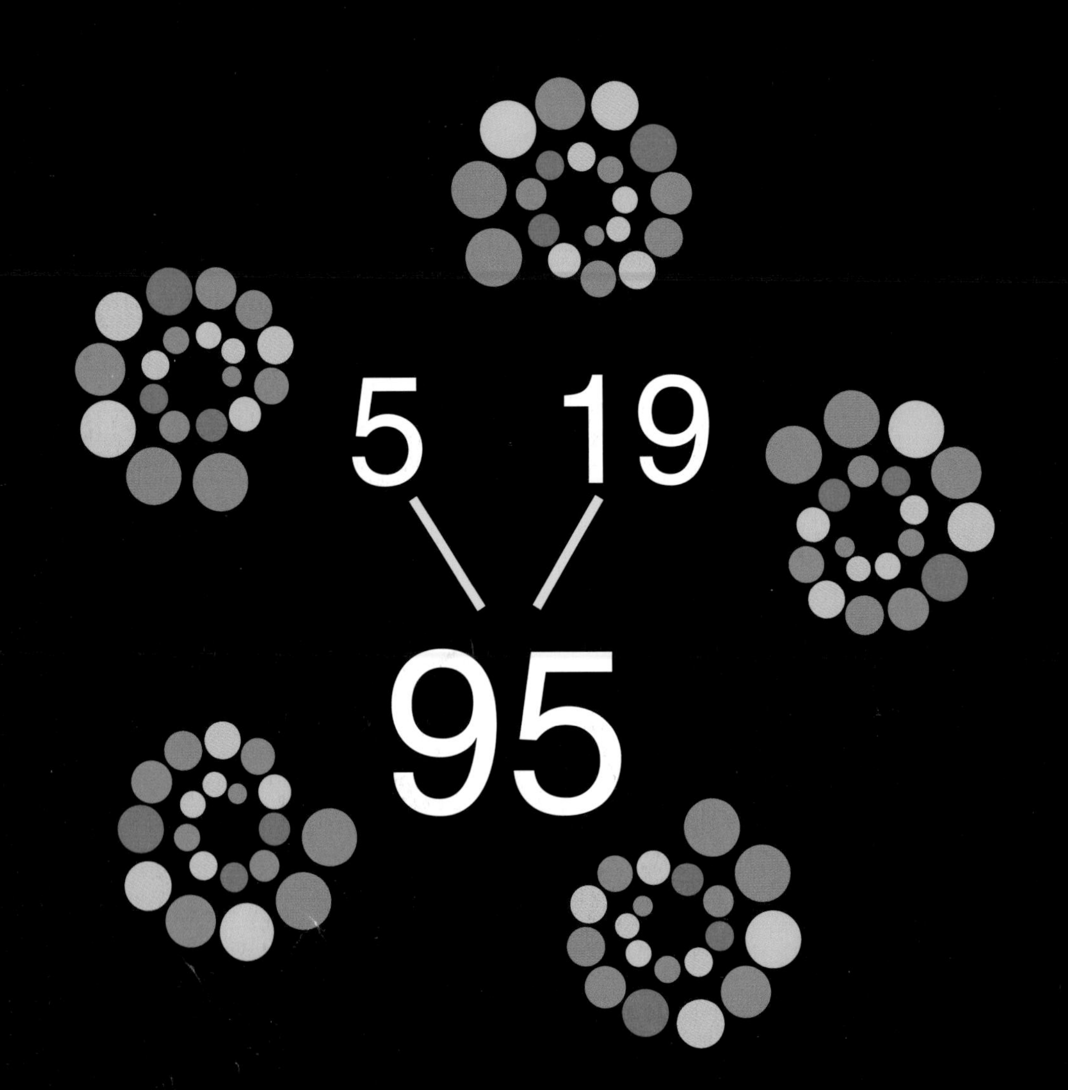
5
19
95

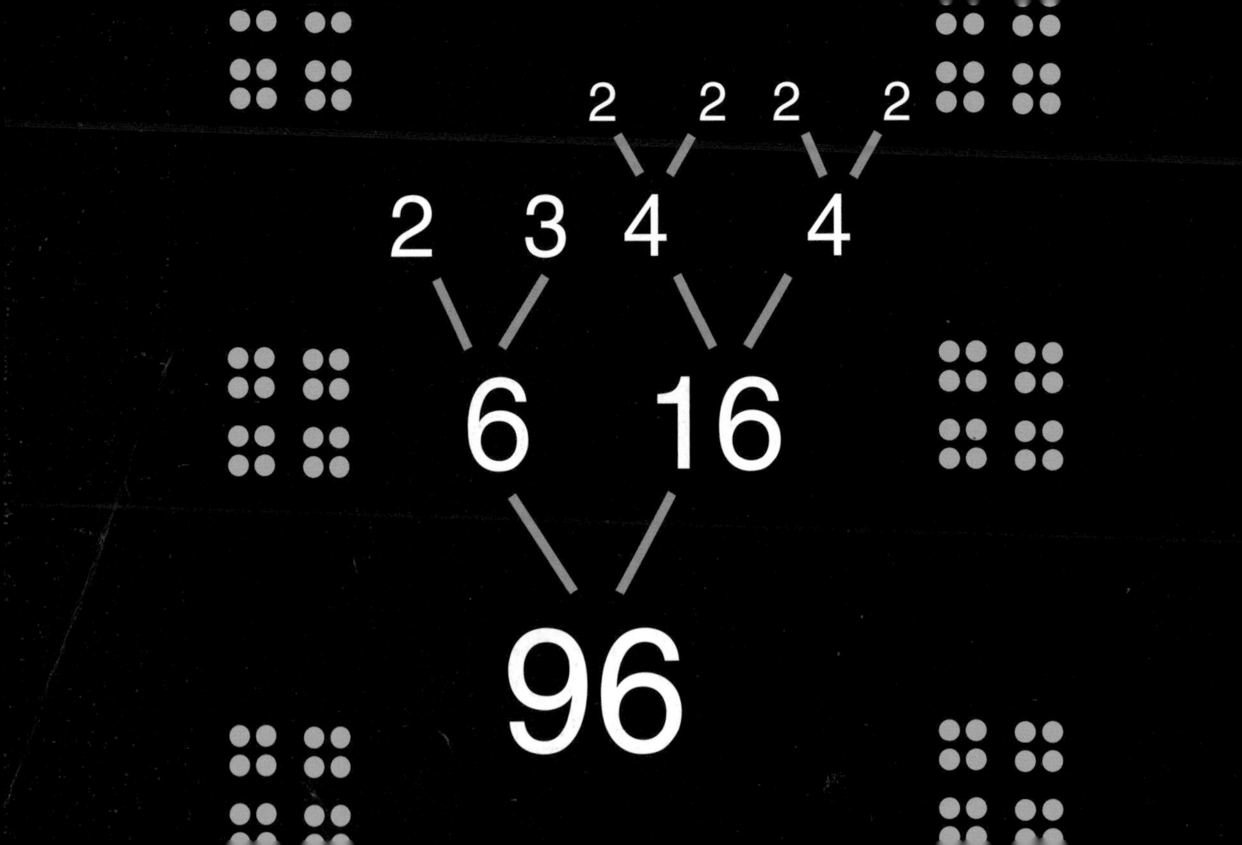

2 2 2 2
2 3 4 4
6 16
96

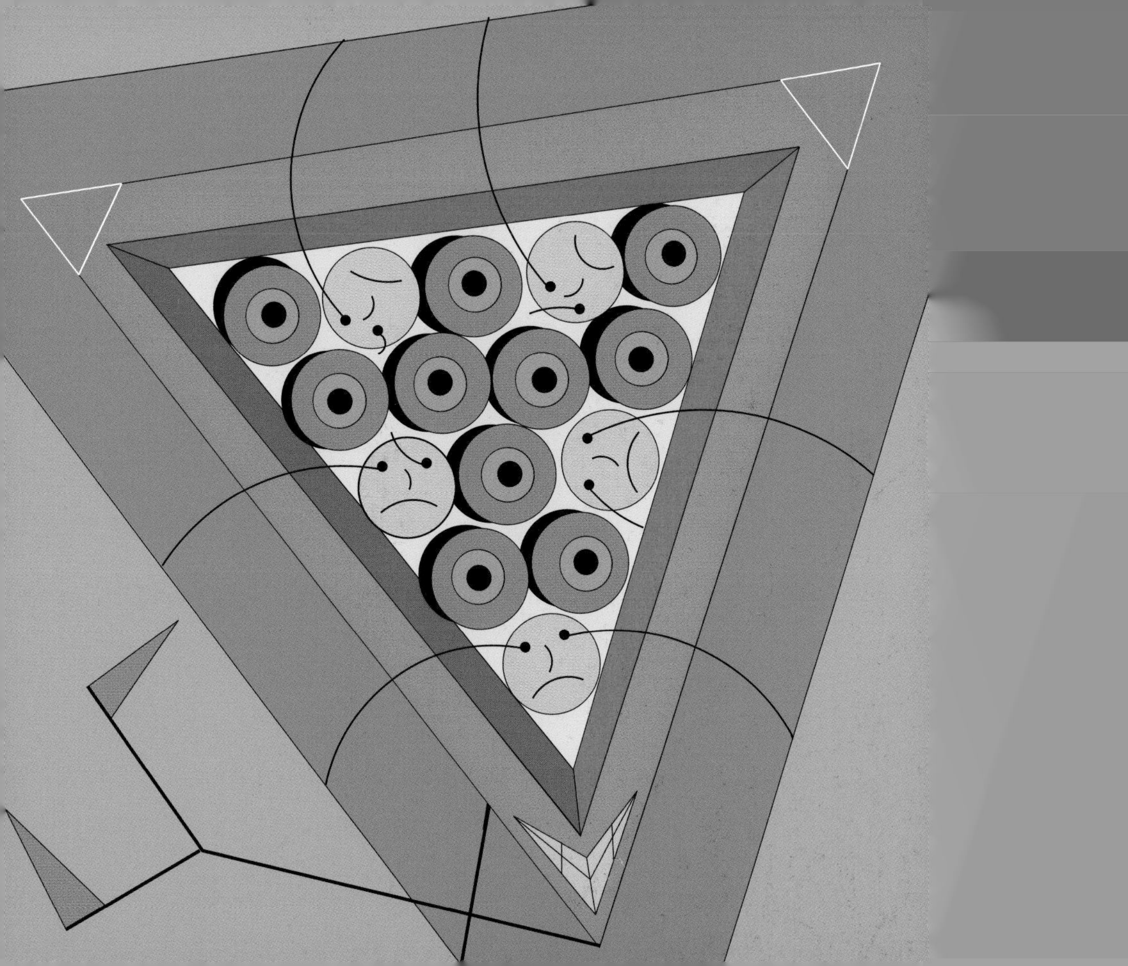

97

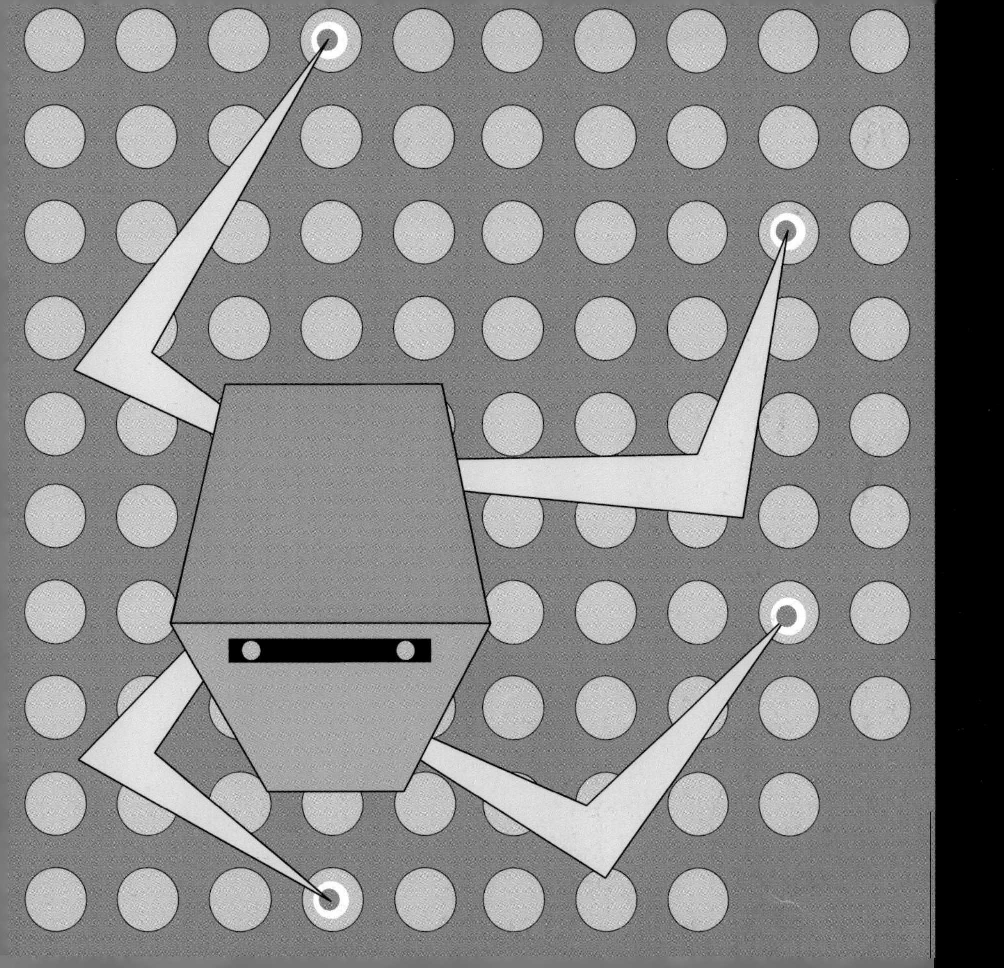

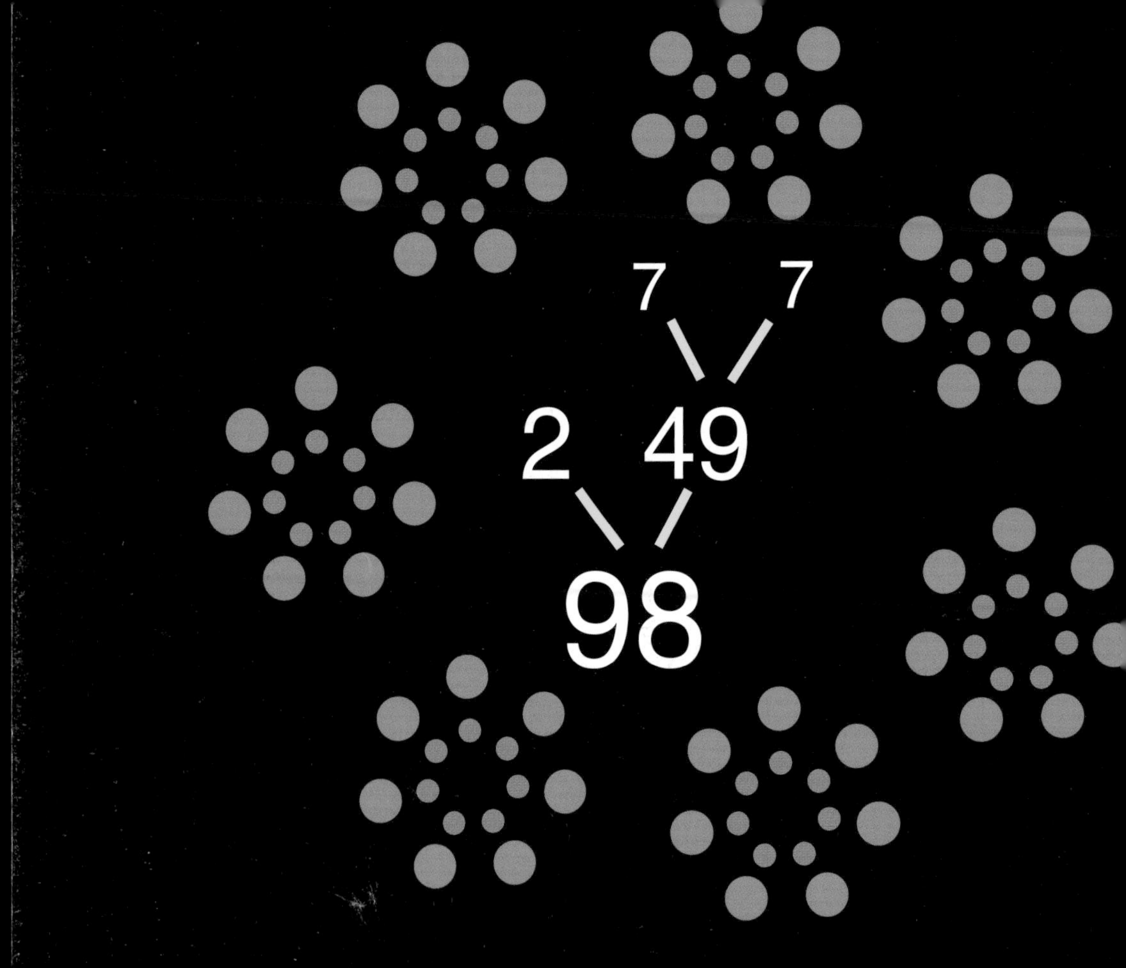
7
7
2
49
98

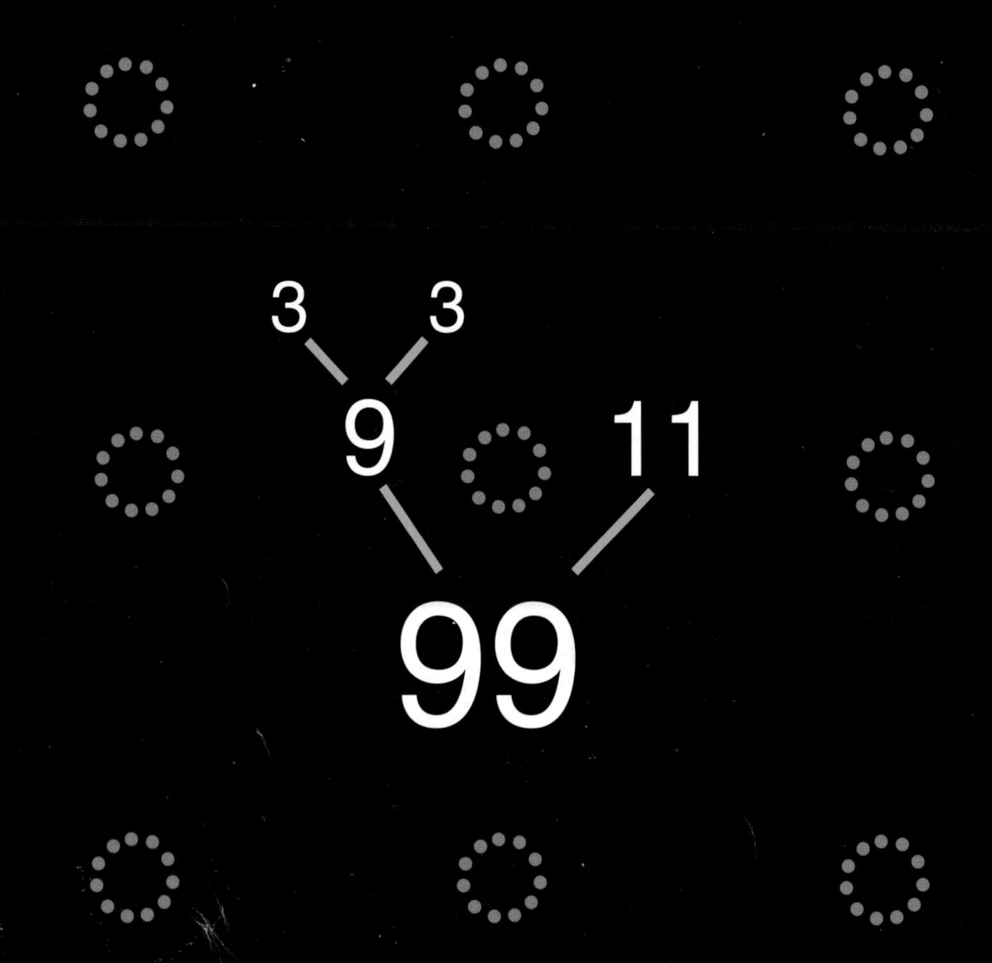
3
3
9
11
99

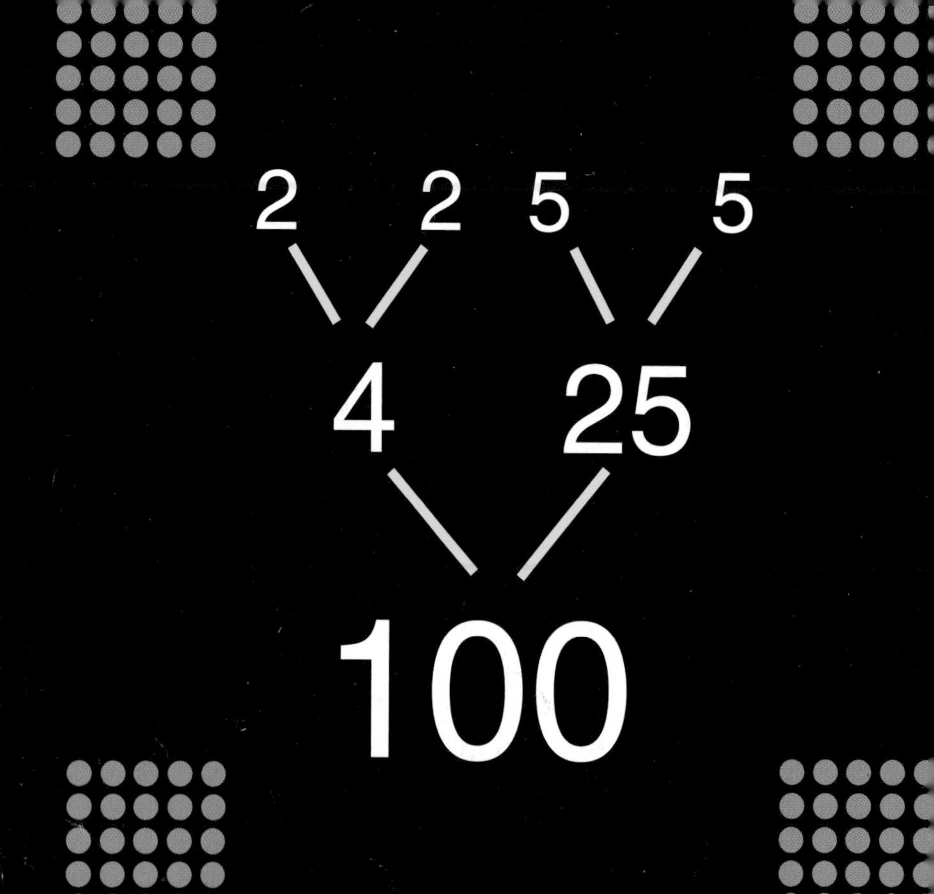
2
2
5
5
4
25
100

How do we
find the primes
less than 100?

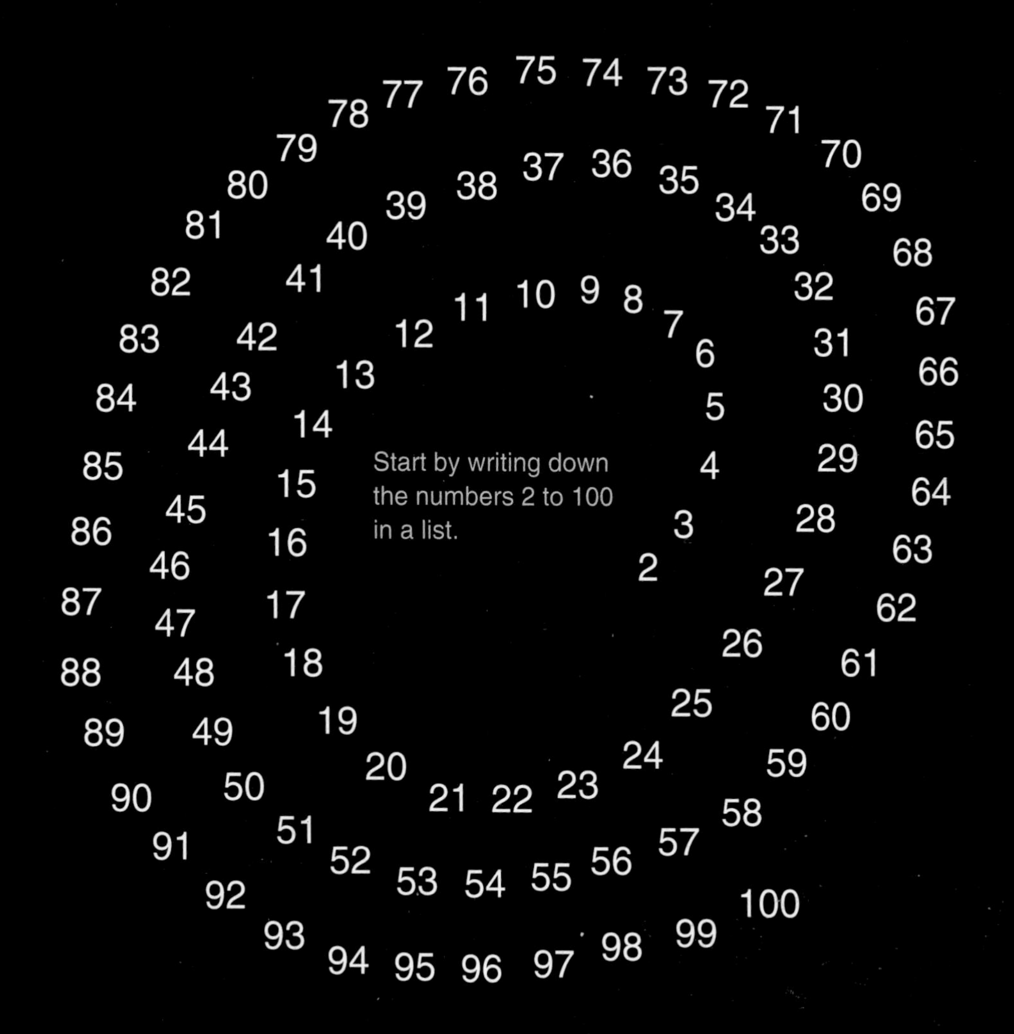
Start by writing down
the numbers 2 to 100
in a list.
2 3 4 5 6 7 8 9 10 11 12 13 14 15 16 17 18 19 20 21 22 23 24 25 26 27 28 29 30 31 32 33 34 35 36 37 38 39 40 41 42 43 44 45 46 47 48 49 50 51 52 53 54 55 56 57 58 59 60 61 62 63 64 65 66 67 68 69 70 71 72 73 74 75 76 77 78 79 80 81 82 83 84 85 86 87 88 89 90 91 92 93 94 95 96 97 98 99 100

Circle the first
number on the
list and cross
off all the
multiples
of that number.

2 3 4 5 6 7 8 9 10 11 12 13 14 15 16 17 18 19 20 21 22 23 24 25 26 27 28 29 30 31 32 33 34 35 36 37 38 39 40 41 42 43 44 45 46 47 48 49 50 51 52 53 54 55 56 57 58 59 60 61 62 63 64 65 66 67 68 69 70 71 72 73 74 75 76 77 78 79 80 81 82 83 84 85 86 87 88 89 90 91 92 93 94 95 96 97 98 99 100

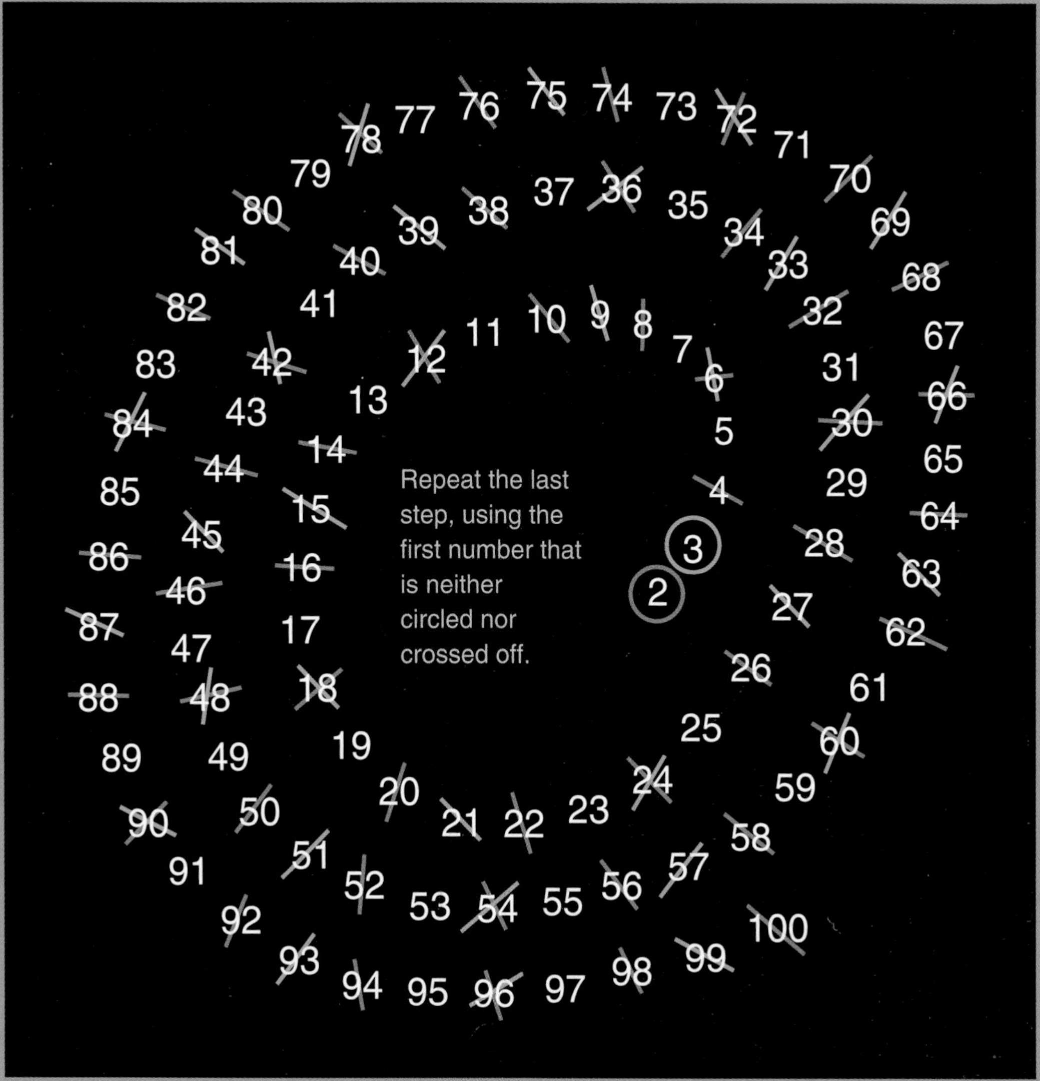

Repeat the last step, using the first number that is neither circled nor crossed off.

Repeat.

2 3 4 5 6 7 8 9 10 11 12 13 14 15 16 17 18 19 20 21 22 23 24 25 26 27 28 29 30 31 32 33 34 35 36 37 38 39 40 41 42 43 44 45 46 47 48 49 50 51 52 53 54 55 56 57 58 59 60 61 62 63 64 65 66 67 68 69 70 71 72 73 74 75 76 77 78 79 80 81 82 83 84 85 86 87 88 89 90 91 92 93 94 95 96 97 98 99 100

Repeat.

2 3 4 5 6 7 8 9 10 11 12 13 14 15 16 17 18 19 20 21 22 23 24 25 26 27 28 29 30 31 32 33 34 35 36 37 38 39 40 41 42 43 44 45 46 47 48 49 50 51 52 53 54 55 56 57 58 59 60 61 62 63 64 65 66 67 68 69 70 71 72 73 74 75 76 77 78 79 80 81 82 83 84 85 86 87 88 89 90 91 92 93 94 95 96 97 98 99 100

Now circle all the
remaining numbers
that are not
crossed out.
The circled
numbers are
the primes less
than 100.

This ancient method is known as the sieve of Eratosthenes. Why can we stop the method after 7? It works because every composite less than 100 is the answer to a multiplication problem involving one of the prime numbers less than 10, and 7 is the largest of these.

You can use the same method to find all the primes less than your favorite number, provided that you have the time and space for it. For example, if you wanted to find all the primes less than 400, you would repeat the basic step until you reached 19. This works because every composite less than 400 is the answer to a multiplication problem involving a prime number less than 20.

Why do the primes
go on forever?

About 2300 years ago, Euclid wrote down an argument proving that the primes go on forever. I'll explain Euclid's argument.

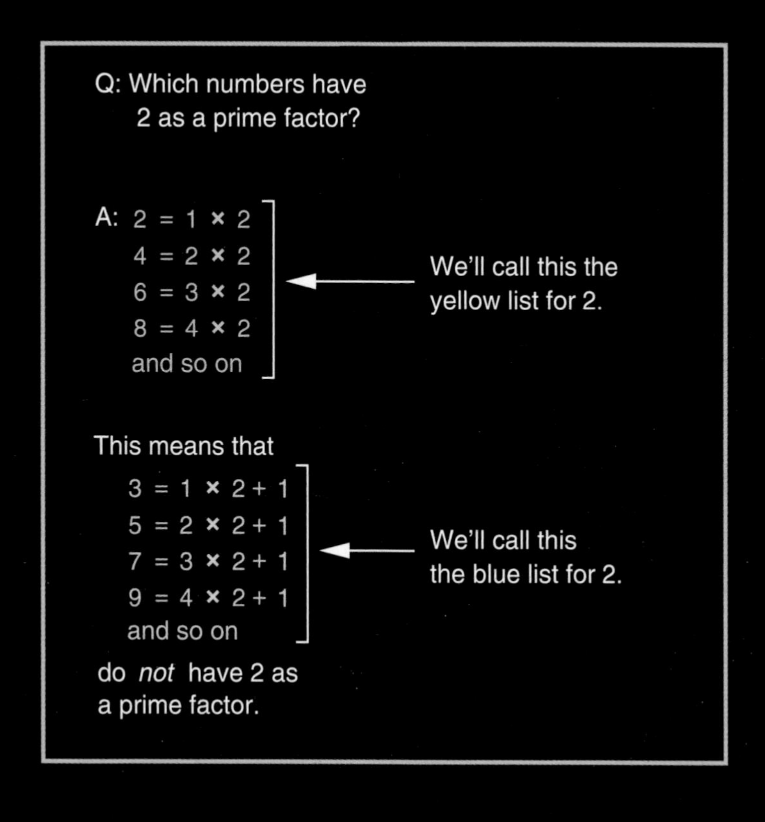

What we just said for 2 can be said for any prime.

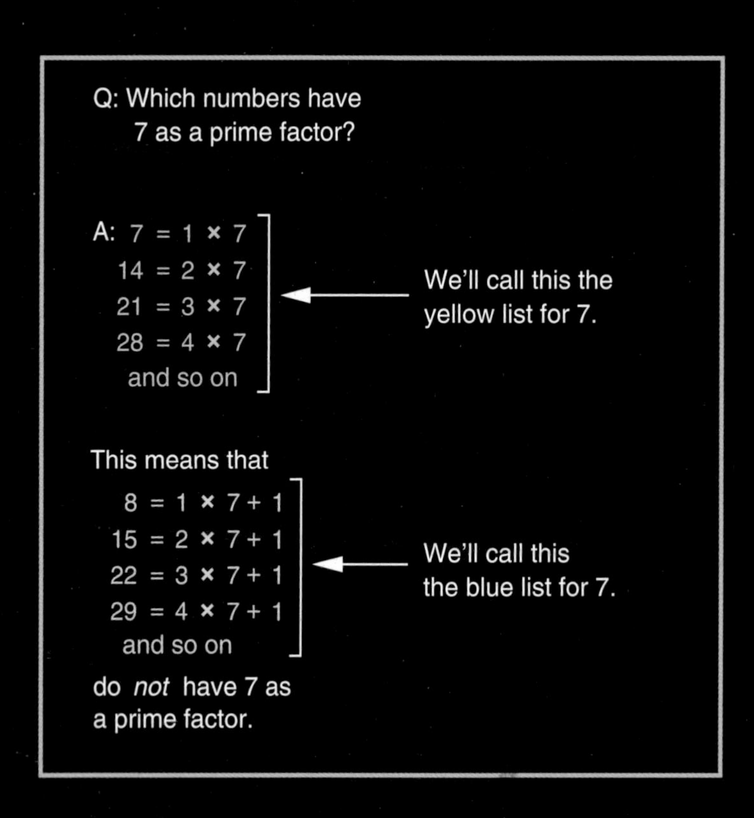

Consider the numbers

$$2! = 2 \times 1 = 2$$

$$3! = 3 \times 2 \times 1 = 6$$

$$4! = 4 \times 3 \times 2 \times 1 = 24$$

$$5! = 5 \times 4 \times 3 \times 2 \times 1 = 120$$

$$6! = 6 \times 5 \times 4 \times 3 \times 2 \times 1 = 720$$

and so on

Let's look at 6!. Think about the primes that are less than 6.

2

6! = 2 × (a whole number)

↓

6! is on the yellow list for 2.

↓

6!+1 is on the blue list for 2.

3

6! = 3 × (a whole number)

↓

6! is on the yellow list for 3.

↓

6!+1 is on the blue list for 3.

5

6! = 5 × (a whole number)

↓

6! is on the yellow list for 5.

↓

6!+1 is on the blue list for 5.

So, the number 6! + 1 does not have 2 or 3 or 5 as a prime factor.

The number 6! + 1 is either prime or composite. Let's consider the possibilities side by side.

If 6! + 1 is prime, then we have found a prime number greater than 6.

If 6! + 1 is composite, then all of its prime factors are greater than 6. We know this thanks to the argument on the previous page. So, again, there is a prime number greater than 6.

All this might look like a lot of work to prove that there is a prime number greater than 6. After all, we already knew about 7 before making this argument. The advantage of this argument is that it works equally well for any other number. Let's watch it in action...

The number 2300! + 1 is either prime or composite.
Let's consider the possibilities side by side

If 2300! + 1 is prime, then we have found a prime number greater than 2300.

If 2300! + 1 is, composite, then all of its prime factors are greater than 2300, because 2300! + 1 is on the blue list for all the primes less than 2300. So, again, there is a prime greater than 2300.

The same argument, done with 2300 in place of 6, shows that there is a prime number greater than 2300. This is something we might not have known in advance of the argument.

The same argument works
for any number. You can pick
any number and know that
there must be a prime
larger than that number.
This means that the primes
go on forever!

ACKNOWLEDGMENTS

I would like to thank my wife, Brienne Brown. Without her encouragement, I never would have seen this project through. I would also like to thank Maureen Stone for her help with the fine-tuning of the colors in this book. Finally, I would like to thank Alice and Klaus Peters for their faith in this book.

ABOUT THE AUTHOR

Richard Evan Schwartz is the Chancellor's Professor of Mathematics at Brown University. He likes creative activities of all sorts, especially drawing cartoonish pictures. For more information about Rich, see http://www.math.brown.edu/~res.